Demystifying Biotechnology

Demystifying Biotechnology

A Comprehensive Guide to Mastering Biotechnology Techniques

Suryaprakash Tripathy

Preface

This comprehensive book, "Demystifying Biotechnology," has been meticulously crafted as an invaluable resource for aspiring bachelor's, master's, and young research scholars in the field of biotechnology. It aims to kindle a passion for exploration and pave the way toward exciting career prospects.

Within these pages, you will find an immersive journey through the day-to-day biophysical and molecular methods utilized in the realms of biochemistry and molecular biology. The book is designed to seamlessly blend theoretical foundations with practical applications, empowering students to make informed decisions and harness techniques effectively. The Virtual learning aids and illustrations serve as beacons of clarity, ensuring an engaging and efficient learning experience.

This book is more than just a guide. It is a catalyst for scientific growth. Alongside the mastery of techniques, I presented scientific challenges with hypothetical scenarios at the end that beckoned readers to design their experiments, fostering a spirit of ingenuity and independent thinking. The chapters in this book are thoughtfully arranged in a logical sequence, with each page meticulously curated to ensure that related text, figures, and tables reside harmoniously together. This book is designed to ensure that users can easily engage with the material without any hassle or inconvenience.

I understand that accessibility is paramount, which is why "Demystifying Biotechnology" is available in eBook, paperback, and hardcover formats.

Acknowledgment

There is an old proverb that says that you never really learn a subject very well until you teach it. I have come to understand the profound truth of that proverb in the process of writing this book. With every word I wrote, I immersed myself further into the world of esteemed literature. I carefully planned to present the subject or concept in a way that is easy for readers to understand, ensuring they grasp the concept thoroughly.

CONTENTS

Abbreviations

µl	Microliter
ABTS	2,2`-azino-bis(3-ethylbenzothiazoline-6-sulfonic acid)
AFLP	Amplified Fragment Length Polymorphism
AFM	Atomic Force Microscope
AH	Anti-Human Globulin
AP	Arbitrary Primer
ATR	Attenuated Total Reflectance
BSA	Bovine Serum Albumin
CD	Circular Dichroism
cDNA	Complementary Deoxyribonucleic Acid
CE	Capillary Electrophoresis
CHFE	Contour-Clamped Homogenous Electric Fields
ChIP	Chromatin Immunoprecipitation
Co-IP	Co-Immunoprecipitation
CPM	Counts Per Minute
Cryo-EM	Cryogenic Electron Microscope
CTAB	Cetyltrimethylammonium Bromide
Da	Dalton
DAPI	4',6-diamidino-2-phenylindole
DAT	Direct Antiglobulin Test
ddNTP	Dideoxynucleotide Triphosphate
DEAE	Diethyl Aminoethyl
DMEM	Dulbecco's Modified Eagle Medium
DMSO	Dimethyl Sulfoxide
DNA	Deoxyribonucleic Acid
dNTP	Deoxynucleotide Triphosphate
DTT	Dithiothreitol
EDTA	Ethylene Diamine Tetra-Acetic acid
EIA	Enzyme Immunoassays
ELISA	Enzyme-Linked Immunosorbent Assay
EtBr	Ethidium Bromide
FBS	Fetal Bovine Serum
FCS	Fetal Calf Serum
FISH	Fluorescence In Situ Hybridization
FITC	Fluorescein Isothiocyanate
FLISA	Fluorescence Linked Immunosorbent Assay
FRET	Fluorescence Resonance Energy Transfer
FTIR	Fourier-Transform Infrared
GC	Gas Chromatography
GITC	Guanidinium Isothiocyanate

GLC	Gas Liquid Chromatography
GLP	Good Lab Practice
GSC	Gas Solid Chromatography
GSP	Gene Specific Primer
HAT	Hypoxanthine-Aminopterin-Thymidine
HEPA	High-Efficiency Particulate Air Filter
HGPRT	Hypoxanthine-Guanine Phosphoribosyl Transferase
HIV	Human Immunodeficiency Virus
HOMO	Highest Occupied Molecular Orbital
HPLC	HIGH-Performance Liquid Chromatography
HRP	Horseradish Peroxidase
IEF	Isoelectric focusing
IFE	Immunofixation Electrophoresis
IP	Immunoprecipitation
IPG	Immobilized pH Gradient
IR	Infrared
IUPAC	International Union of Pure and Applied Chemistry
LC	Liquid Chromatography
LUMO	Lowest Unoccupied Molecular Orbital
mAb	Monoclonal Antibodies
Mb	Metabases
MEM	Minimum Essential Medium
ml	Milliliter
MMR	Measles, Mumps, and Rubella
MS	Mass Spectrometry
MSDS	Material Safety Data Sheets
NA	Numerical Aperture
NMR	Nuclear Magnetic Resonance
OECD	Organization for Economic Co-operation and Development
ORD	Optical Rotatory Dispersion
PAGE	Polyacrylamide Gel Electrophoresis
PBS	Phosphate-Buffered Saline
PC	Paper Chromatography
PCR	Polymerase Chain Reaction
PEG	Polyethylene Glycol
pH	Potential of Hydrogen
pI	Isoelectric Point
PPB	Parts Per Billion
PPE	Personal Protective Equipment
PPM	Parts Per Million
PTM	Post-Translational Modification
PVDF	Polyvinylidene Fluoride

PVP	Polyvinylpyrrolidone
qRT PCR	Quantitative Real-Time Polymerase Chain Reaction
RACE	Rapid Amplification of cDNA Ends
RAPD	Rapid Amplification of Polymorphic DNA
RBC	Red Blood Corpuscles
RCF	Relative Centrifugal Field
Rf	Relative Front
RIA	Radioimmunoassay
RIP	RNA Immunoprecipitation
RIPA	Radioimmunoprecipitation Assay
RNA	Ribonucleic Acid
RPM	Revolutions Per Minute
RPMI	Roswell Park Memorial Institute
RT PCR	Reverse Transcription Polymerase Chain Reaction
SDS	Sodium Dodecyl Sulfate
SEC	Size Exclusion Chromatography
SEM	Scanning Electron Microscope
SIM	Structured Illumination Microscopy
SNR	Signal-to-Noise Ratio
SPM	Scanning Probe Microscope
SSC	Saline Sodium Citrate
STED	Stimulated Emission Depletion
STM	Scanning Tunnelling Microscope
TAE	Tris Acetate EDTA
TBE	Tris Borate EDTA
TBST	Tris-buffered saline Tween 20
TBST	Tris-Buffered Saline Tween 20
TEM	Transmission Electron Microscope
TEMED	Tetramethyl Ethylenediamine
TLC	Thin Layer Chromatography
TMB	3,3`,5,5`-Tetramethylbenzidine
TMS	Tetramethylsilane
UV	Ultraviolet
Vis	Visible
Y2H	Yeast Two-Hybrid

Introduction

Before we dive into this exciting learning process, let me introduce you to the unique and futuristic aspects of this book. It is designed to make your learning experience not just informative but also engaging. As you progress, you will not only grasp biotechnology concepts but also enhance your analytical and critical thinking skills which will help you to secure a prestigious position in biotechnology and related fields.

Perk-1

The information within this book comes from trusted sources, including experts in their respective fields and the latest review papers. This ensures that the data presented in the book is accurate and reliable.

Perk-2

This book is packed with helpful visuals, tables, analogies, examples, and real-world scenarios. These tools make it a breeze to understand even the trickiest concepts. Plus, they will ignite your passion for learning biotechnology techniques, making the journey even more exciting!

Perk-3

This book does not just give you the concepts and step-by-step instructions for each technique; it goes the extra mile. Inside, you will find a special section called 'Virtual Learning Aid' where you can access high-quality videos demonstrating lab techniques and animations. These resources are designed to make understanding the practical side of biotechnology even more accessible and engaging for you.

Note- I invite our readers to enhance their experience by downloading a suitable scanner app to unlock the exciting content hidden behind QR codes. The selection of videos is based solely on their excellent teaching quality and engaging lab demonstrations, with no bias involved.

Perk-4

As you read, you will discover a clever way I have used to make important concepts stand out. This means if you are already familiar with a concept, you can easily skip the details and go through the highlighted portions and keep your reading experience smooth and enjoyable.

Perk-5

Inside this book, you will find a special chapter called 'Designing Experiments' that is all about making learning fun and practical. In this chapter, you will step into hypothetical scenarios that let you apply what you have learned. It's like a training ground where you will tackle exciting research challenges faced by scientists and students in labs. By working through these scenarios, you will sharpen your critical thinking skills, making you stand out in the research world and contribute meaningfully to the scientific community.

Note- This chapter owes its creation to the remarkable power of Artificial Intelligence. It has enabled me to craft imaginative scenarios by combining various techniques and scientific approaches.

Perk-6

I warmly invite readers to receive a free copy of my upcoming books by sharing your futuristic suggestions for further improvement of the book! In my upcoming books, I will give special recognition to readers who provide valuable suggestions that enhance the quality of content and improve the overall reading experience. Your name will be featured in a dedicated section with heartfelt gratitude from both the author and fellow readers.

Note: Readers can connect with author by visiting the "Author's Insight" section at the end of this book (**Page 359**).

Chapter 1

Good Lab Practices

Introduction

Good Lab Practices (GLP) encompass a comprehensive set of principles that govern various aspects of laboratory work, ensuring scientific data's integrity, reliability, and traceability. It covers all stages of research, including **experimental design**, **execution**, **data management**, and **reporting**.

To better grasp the concept, let's imagine an analogy with the art of cooking. Imagine you are aiming to create a mouthwatering dish. As an experienced cook, you know that following a well-crafted recipe is vital to achieve the desired outcome. Just as a recipe provides a step-by-step guide, **GLP offers a structured approach for researchers to plan their experiments**, ensuring that all necessary variables are considered. Just as a chef pays attention to ingredient quality, **scientists must select reliable materials and equipment**. Just as a cook maintains a clean and organized kitchen, the **researcher must maintain a sterile and well-maintained laboratory environment**. Finally, just as a chef records the ingredients used and cooking times, **scientists must diligently document their procedures, observations, and results**.

Throughout history, the absence of proper lab practices has led to several catastrophic incidents and erroneous conclusions. Consider the analogy of a kitchen where a chef disregards recipe instruction, fails to maintain cleanliness, or overlooks the importance of precise measurements. Such negligence often results in a dish that is inconsistent or even dangerous to consume. Similarly, in the laboratory, **a lack of adherence to GLP can lead to flawed experiments, unreliable data, and potentially hazardous outcomes**. For example, In the late 1990s, Andrew Wakefield, a British doctor, published a study suggesting a link between the Measles, Mumps, and Rubella (MMR) vaccine and autism. It was later found to have serious flaws and ethical misconduct. The study had a small sample size, lacked proper control groups, and relied on subjective data collection. Wakefield's actions led to a decline in Vaccination rates and outbreaks of preventable diseases. This example serves as a cautionary tale, illustrating how

Poor lab practices can have critical consequences. It underscores the critical need for rigorous adherence to GLP, including **proper study design, robust data collection, ethical considerations**, and **accurate reporting**. By following these instructions, we can **ensure the reliability and credibility of research, safeguard public health,** and **prevent the spread of misinformation**.

New Zealand and Denmark were the first to introduce GLP in 1972. United States was the next to introduce GLP in response to the poor scientific practices prevalent in the US around that time. The **Organisation for Economic Co-operation and Development (OECD) published the principle of GLP** in 1981 under the name **"OECD Guidelines for Testing Chemicals"** that were internationally accepted.

Virtual learning aid

Guidelines and precautions

To implement laboratory protocol, it is essential to focus on specific guidelines from the OECD frameworks, which are carefully selected and mentioned below.
1. **Personal protective equipment (PPE) is crucial for ensuring the safety of laboratory personnel**. Wearing an **apron** protects clothing from spills and splashes. **Gloves** protect hands from chemical exposure and potential contamination. **Safety glasses** shield the eyes from hazardous materials or flying particles, and proper **footwear** protects against spills and falling objects. Adhering to PPE guidelines minimizes the risk of injuries and exposure to harmful substances.

2. **It is essential to carefully read and understand the experiment's procedure before conducting any laboratory work**. This allows you to **familiarize yourself with the steps involved, anticipate potential hazards,** and ensure that you have the necessary materials and equipment ready. Understanding the procedure beforehand helps **minimize errors, promotes accurate data collection,** and **enhances overall experimental efficiency**.

3. **Proper labeling of materials is critical for maintaining organization and avoiding mix-ups**. Labeling ensures that you can **accurately identify** and **track the contents** of each container, **preventing errors and confusion**. Clear and legible labels also contribute to **proper documentation** and **traceability on experiments.**

4. **Before using any substance or reagent, always verify its identity and check for any specific handling instructions or warnings indicated on the label**. This practice helps **prevent accidental use** of the wrong substance, ensures **correct measurements**, and **minimizes the risk of hazardous reactions**.

5. **Familiarize yourself with the locations and proper usage of safety equipment in the laboratory**. This includes knowing the nearest **safety showers** and **eyewash stations** in case of chemical splashes, the location of **fire extinguishers** for immediate response to fires, and the availability and accessibility of **first aid kits** for addressing injuries or accidents. Being aware about the safety equipment and its procedures ensures swift and appropriate action during emergencies.

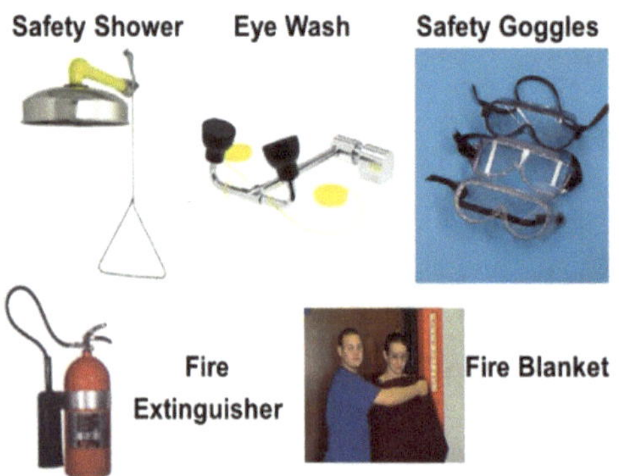

6. **Chemical fume hoods are designed to provide a controlled environment for handling hazardous substances**. When working **with strong acids**, **fuming chemicals**, or **volatile solvents**, it is important to utilize a chemical fume hood. The fume hood helps contain and exhaust harmful vapors, protecting you from inhalation and **minimizing the risk** of exposure to hazardous substances.

7. **Microorganisms can pose various risks, including infections and the potential for contamination**. Depending on the nature and level of risk associated with the microorganisms, **specific biosafety-level containment**

protocols must be followed. These protocols provide guidelines for **handling**, **storage**, and **disposal of microorganisms**, ensuring the **safety** of laboratory personnel and **preventing the spread** of infectious agents.

Comparison between Biosafety levels and their possible risks.

Biosafety Level (BSL)	Possible Risks	Example
BSL-1	Low risk; minimal potential harm to humans or the environment	*Escherichia coli* (non-pathogenic strains), *Saccharomyces cerevisiae* (baker's yeast)
BSL-2	Moderate risk; can cause disease in humans, but treatments are available	*Staphylococcus aureus*, Salmonella species, Hepatitis B virus
BSL-3	High risk; can cause serious or potentially lethal diseases, often no vaccines available	*Mycobacterium tuberculosis*, Yersinia pestis (plague), HIV
BSL-4	Maximum risk; high lethality, no vaccines or treatments available	Ebola virus, Marburg virus, Variola virus (smallpox)

8. **Maintaining a clean and decontaminated work area is essential for preventing cross-contamination and ensuring the integrity for experimental results**. After completing an experiment or handling hazardous materials, it is important to **properly decontaminate** your workbench and any equipment used. This typically involves cleaning surfaces with appropriate disinfectants or decontamination agents, ensuring **a sterile environment** for subsequent experiments.

9. **When working with infectious materials, it is crucial to follow proper decontamination procedures before disposal**. This may involve **sterilization methods** such as **autoclaving** or using specific **disinfectants**. Decontaminating infectious materials before disposal minimizes the risk on spreading pathogens and **ensures compliance** with biosafety regulations.

Difference between Sanitization, Disinfection, and Sterilization.

Property	Sanitization	Disinfection	Sterilization
Purpose	Reduce germs	Kill most germs	Eliminate all germs
Reagents	Soap and water	Disinfectants	Heat, Chemicals, or Radiation
Examples	Hand washing	Surface cleaning	Surgical tools, laboratory equipment
Effect on Spores	Does not kill	May kill some spores	Kills most spores
Residue	Minimal residue	May leave some residue	May leave some residue
Applicability	Everyday use	Healthcare settings, etc.	Medical, laboratories

10. **In the event of a spill, accident, or injury, it is essential to report the incident immediately to your instructor or the appropriate authority**. Prompt reporting allows for timely response and necessary actions **to mitigate potential risks**, **provide proper medical attention** if required, and **prevent further incidents**. Reporting incidents also aids in **identifying potential hazards** and **improving laboratory safety** protocols.

In addition to adhering to OECD guidelines, it is vital to utilize **Material Safety Data Sheets (MSDS)** for ensuring laboratory safety. MSDS is a document provided **by the manufacturer or supplier** that contains essential information **about the potential hazards of a material**. It includes sections on **product information, ingredient details, physical properties, fire and explosion hazards, toxicity data, first aid measures**, and guidelines for **safe handling, usage, storage**, and **disposal**. MSDS also employs **pictograms** on chemical containers, allowing for **quick identification** of material hazards.

These sheets play a crucial role in enhancing both scientific understanding and reader-friendly communication of material safety in the laboratory.

Prohibited actions

1. **Laboratory benches are designated workspaces meant for experiments and should not be used as seating**. Sitting on laboratory benches can **introduce contaminants** or **interfere with ongoing experiments**, compromising safety and accuracy.

2. **Consuming food or beverages inside the laboratory poses a significant risk of contamination**. Accidental ingestion of hazardous substances can have serious health consequences. To maintain a safe working environment, it is crucial to strictly **avoid eating** or **drinking** in the lab.

3. **Contact lenses and cosmetics should not be handled or applied in the laboratory**. Chemicals or substances present in the lab may interact with contact lenses or contaminate cosmetics, leading to **eye irritation**, **infections**, or other adverse reactions. It is important to **prioritize safety** by refraining from these practices in the lab.

4. **Chemicals and reagents used in the laboratory can be hazardous and potentially harmful to the skin**. It is essential to always wear appropriate **gloves** when handling chemicals to protect against potential skin contact and absorption of harmful substances.

5. **Attempting to test or smell chemicals directly poses unnecessary risks**. Certain chemicals may be **toxic**, **corrosive**, or **reactive**, and direct contact or inhalation can result in **harmful effects**. It is important to rely on established testing methods and follow safety protocols when dealing with chemicals.

6. **Pipetting chemicals by mouth is an extremely dangerous practice and should never be done**. It can lead to **ingestion** of toxic or harmful substances, accidental **aspiration**, and the **spread of contaminants**. Proper pipetting techniques, using appropriate equipment such as pipettes, should always be followed.

7. **Returning excess chemicals to the original container can potentially contaminate the entire supply**. To maintain the quality of chemicals, it is important to use **separate containers for excess amounts** and follow **proper disposal** procedures for unused substances.

8. **Diluting concentrated acids requires caution to prevent hazardous reactions**. Adding water to concentrated acids directly can result **in rapid heat generation** and **splattering**, causing **burns** or **explosions**. The recommended procedure is to **add concentrated acids slowly to the water** while stirring continuously to ensure **proper mixing** and heat dissipation.

9. **It is essential to always monitor ongoing experiments and never leave them unattended**. Unattended experiments can lead to **uncontrolled reactions, equipment failures**, or potential **safety hazards**. Being present and attentive during experiments allows for prompt action and intervention in case needed.

10. **When handling glassware or vials, it is important to exercise caution and avoid pointing the open end towards yourself or others**. Accidental **spills, splashes**, or **breakage** can result in **injuries**. Maintaining proper handling techniques, such as holding glassware away from the body and others, helps **minimize the risk** of accidents.

Laboratory notebook

Lab notebooks serve as vital tools for **recording experimental procedures, observations, data**, and **analysis**. In this section, we will explore the significance of maintaining well-organized and comprehensive laboratory notebooks and how they contribute to the integrity and progress of scientific research.

All researchers or students experimenting in the lab should **prepare written reports of the results of experimental work**. Since this record may be studied by many individuals, it must be completed, in a clear, concise, orderly, and accurate manner. It should have the below details in a clear and organized manner.

Introduction

      a. Objective or purpose
      b. Theory

The introduction section of laboratory notebooks plays a crucial role in providing a concise and focused overview of the experiment's **objective or purpose**. It should begin with a three- or four-sentence statement that clearly **outlines the goals or aims of the experiment**. To formulate this statement, it is helpful to ask yourself questions such as "What are the specific objectives I aim to achieve through this experiment?"

Following this objective statement, a brief discussion of the underlying **theory behind the experiment** should be included, providing the necessary **context** and **background information**. This discussion helps readers understand the scientific principles and concepts that form the basis of the experiment, ensuring a comprehensive understanding of its purpose and relevance.

Experimental

      a) Tables of materials and reagents
      b) List of Equipment
      c) Flowchart
      d) Record of procedure

The Experimental section of laboratory notebooks commences with a **comprehensive list of all the reagents and materials** employed in the experiment. It is important to include each chemical's source and specify the solutions' concentrations. For instrumentation, **providing references** to the **company name** and **model number** ensures accurate identification and traceability. A stepwise procedure for the experiment can be depicted using a **flowchart** to enhance clarity and ease of understanding. This visual representation allows for a clear and logical presentation of the experimental steps, complementing the list of Equipment and facilitating a reader-friendly approach to following the experiment.

Data and Calculations

      a) Recording of all raw data including printouts
      b) Method of calculation with statistical analysis
      c) Present final data in tables, graphs, or figures when appropriate

In the Data and Calculation section of your laboratory notebook, it is essential to **record all raw data** directly in the **notebook** itself rather than using separate sheets or paper towels. This ensures that all data is kept together and prevents the **risk of losing or misplacing valuable information**. Additionally, proper **statistical analysis** should be incorporated in this section to provide a scientifically rigorous approach to **analyzing and interpreting the collected data**. This includes applying appropriate statistical methods to draw meaningful conclusions from the experimental results and enhancing the reliability and validity of the findings.

Results and Discussion

    a) Conclusions
    b) Compare the result with known values
    c) Discuss the significance of the data
    d) Literature reference

The Results and Discussion section of your laboratory notebook holds significant importance as it provides **answers to the fundamental questions** of whether you achieved your proposed goals and objectives. Any conclusions drawn must be **supported by the results obtained from your experiments**. It is valuable to compare your data with known values and existing literature, if applicable, and calculate the percentage error **to explain any discrepancies**. Additionally, if you encounter any challenges during the experiment, it is crucial to outline them and **suggest possible remedies for future experiments**. Finally, make sure to include a list of all the **references from books, journal articles**, and **websites** that were utilized to compile the experiment write-up. This section helps **ensure the scientific integrity** of your work and allows readers to explore further resources for a deeper understanding of the experiment.

Chapter 2

Operation Of Common Laboratory Instruments

Introduction

In this chapter, we will discuss commonly used **laboratory instruments** and **procedures**, also we will demystify **their importance in the biotechnology landscape**.

Liquid handling system

Liquid handling systems are sophisticated instruments designed to precisely **manipulate and transfer liquids** in scientific experiments. They play a pivotal role in various fields, including biotechnology, pharmaceutical research, and molecular biology. It can be broadly categorized into three types:

1. **Disposable Pasteur Pipets**: Disposable Pasteur pipets are commonly used when a **semi-quantitative transfer** of a **small volume**, typically ranging from **1 to 10 ml**, needs to be performed. These pipets are ideal for situations **where precise measurements are not critical**, but a general transfer of liquid is required. They are often made of **plastic** or **glass** and feature **a long, slender shape with a narrowed tip** for easy aspiration and dispensing of liquids.

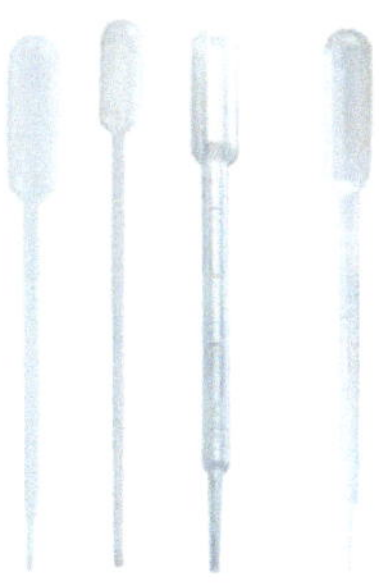

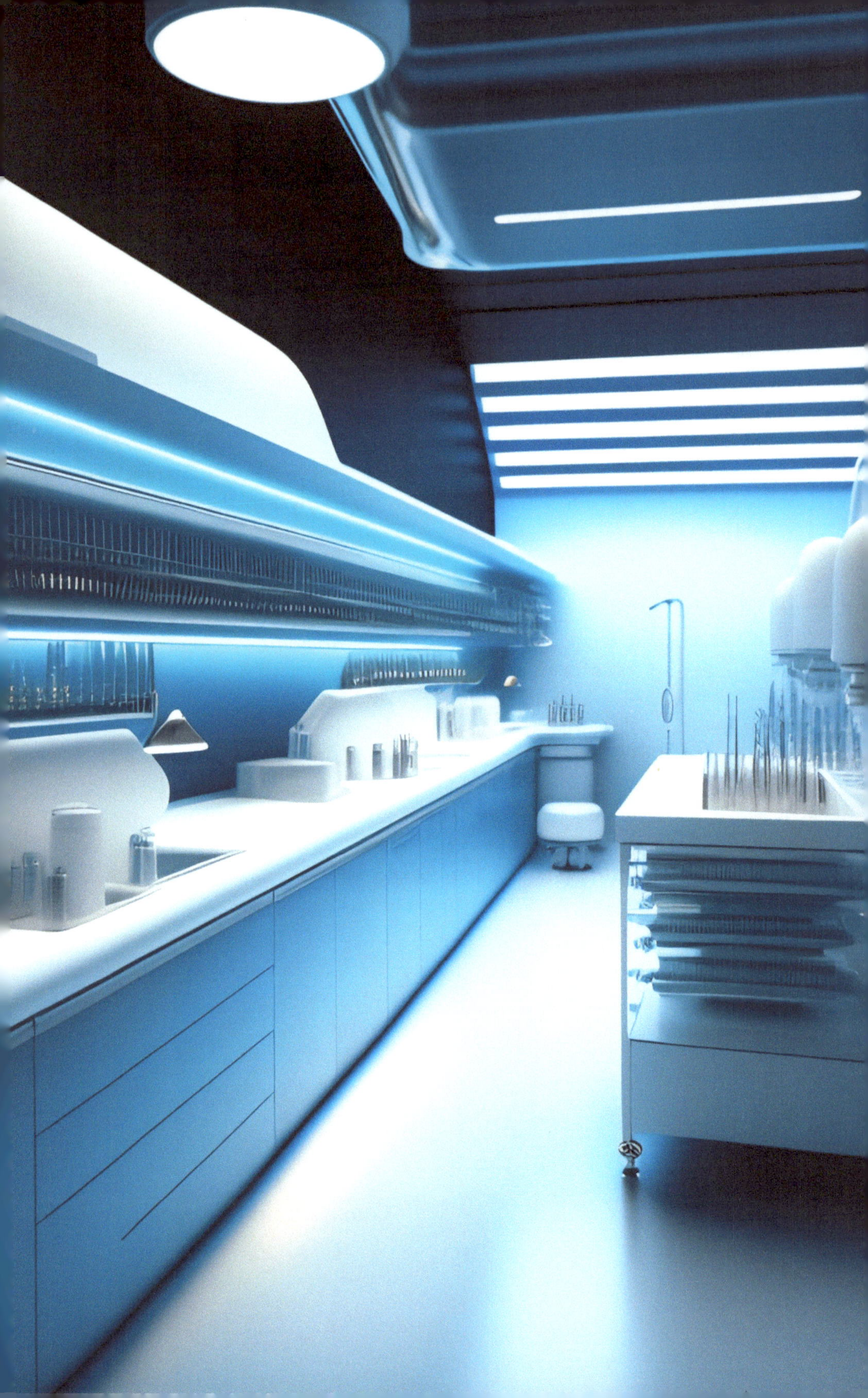

2. **Calibrated Pipets**: When a **quantitative transfer** of a specific and **accurate volume** is required, calibrated pipets come into play. These pipets are designed to **deliver precise volumes** of liquids. They are **available in various volume ranges**, allowing scientists to select the appropriate pipet for their specific needs. Calibrated pipets undergo meticulous calibration processes to ensure accuracy, typically using gravimetric or volumetric techniques. They are commonly used in analytical laboratories, where precise measurements are crucial for obtaining reliable data.

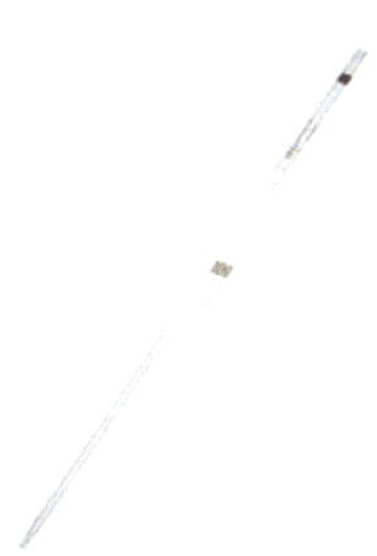

3. **Automated Pipetting Systems**: For situations that **involve multiple identical small-volume transfers** or require rapid and precise dispensing, automated pipetting systems are the ideal choice. These systems utilize mechanical microliter pipettors, which are capable of **accurate**, **precise**, and **rapid dispensing** of fixed volumes ranging from **1 to 5000 μl (5 ml)**. Automated pipetting systems offer advanced features such as programmability, multiple dispensing modes, and sample tracking capabilities. They significantly improve laboratory productivity by reducing manual pipetting errors and saving time on repetitive tasks.

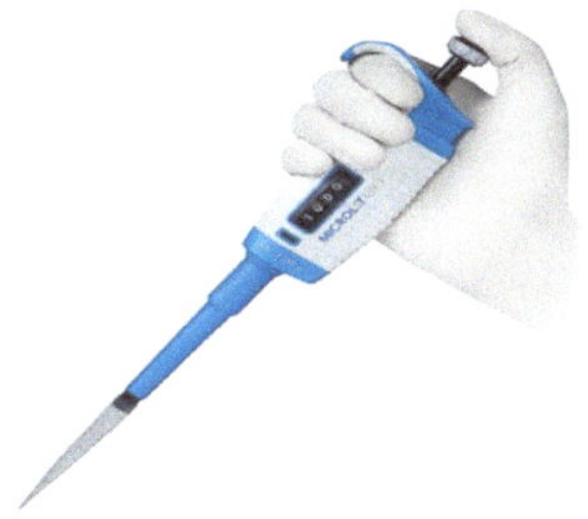

Protocol

1. **Familiarize Yourself**: Begin by thoroughly familiarizing yourself with the specific model and user manual of the **automated pipette you are using**. Understand its features, functionalities, and any specific operating instructions.

2. **Proper Handling**: Always handle the automated pipette with care. **Hold it by the handle or grip provided**, avoiding contact with sensitive components or the dispensing tip.

3. **Calibration Check**: Regularly perform calibration checks according to the **manufacturer's recommendations** to ensure accurate and precise pipetting. Follow the calibration procedure provided in the user manual.

4. **Proper Tip Selection**: Use appropriate pipette tips that are **compatible** with the automated pipette model. Ensure that the **tips are securely attached to prevent leakage or air gaps**.

5. **Pre-Wetting the Tip**: Before aspirating the sample, pre-wet the pipette tip by dispensing and discarding the liquid **without touching the sides of the container**. This helps **to ensure accurate volume measurements**.

6. **Sample Aspiration**: Immerse the pipette tip into the sample solution, **ensuring that it is fully submerged**. Depress the plunger button smoothly and steadily to aspirate the desired volume.

7. **Dispensing Technique**: Position the pipette tip just above the target container and **depress the plunger button gently and uniformly to dispense the sample. Release the button slowly** to prevent any liquid from being drawn back into the tip.

8. **Tip Ejection**: After dispensing, **carefully eject** the used tip into a designated **waste container**. Ensure that the tip is completely detached from the pipette to avoid any cross-contamination.

9. **Cleaning and Maintenance**: Regularly clean the external surfaces of the automated pipette **with a mild disinfectant solution**. Pay attention to the dispensing mechanism and ensure that it is free from any residues or blockages. Follow the **manufacturer's guidelines** for maintenance and servicing.

10. **Storage**: After use, **store** the automated pipette **in a designated location**, following any specific storage instructions provided by the manufacturer. **Protect it from dust, moisture,** and **extreme temperatures**.

pH meter

A pH meter is a scientific instrument used to measure the **acidity or alkalinity of a solution**. It **determines the concentration of hydrogen ions (H+) in the solution**, which is a measure of its **pH value**. The pH meter consists of a **glass electrode** and **a reference electrode**, both **immersed in the solution** being tested. The electrodes **generate a small electrical potential** that is **converted into a pH reading** on the meter's display. The significance of a pH meter in the laboratory is immense. pH is a fundamental property of solutions and plays a crucial role in various scientific disciplines, including chemistry, biology, environmental science, and medicine.

Here are a few key aspects of the significance of pH meters:

1. **Chemical Analysis**: pH measurement is essential for determining the **acidity** or **alkalinity** of chemical solutions. It helps in quantifying acids or bases present, enabling scientists to monitor and control chemical reactions accurately. pH meters are widely used in **titrations**, **quality control analysis**, and the **preparation of chemical solutions**.

2. **Biological and Environmental Studies**: **pH** is a critical factor in biological systems and environmental studies. It **affects enzyme activity**, **microbial growth**, and the **physiological processes** of living organisms. pH meters are invaluable tools in fields such as **biochemistry, microbiology, agriculture**, and **environmental science**. They help researchers monitor pH levels in biological samples, soils, water bodies, and various environmental compartments, providing insights into ecosystems and their health.

3. **Medical and Pharmaceutical Applications**: **pH** measurement is **vital in medical diagnostics, pharmaceutical research**, and **drug formulation**. In medical diagnostics, pH meters are used **to analyze bodily fluids**, such as blood, urine, and saliva, **to assess health conditions** and **diagnose diseases**.

4. **Industrial Processes**: **pH** control is essential in numerous industrial processes, such as **food and beverage production, water treatment**, and **chemical manufacturing**. pH meters are employed to monitor and regulate pH levels in these processes, **ensuring product quality, process efficiency**, and **regulatory compliance**.

How pH meter measure pH?

pH measurement requires **two electrodes**:
1. A **pH-dependent electrode** sensitive to H^+
2. A **pH-independent calomel reference electrode**.

The potential difference between two electrodes is measured as a voltage by utilizing a formula given below:

$$V = Econstant + \frac{2.303RT}{F}pH$$

Where,
V = Voltage of the complete circuit
Econstant = Potential of the reference electrode
R = Gas constant
T = Temperature
F = Faraday constant

Virtual learning aid

Protocol

1. **Prepare the pH meter**: Ensure that the pH meter is **clean** and in **good working condition** before use. Check the electrode for any signs of damage or contamination. If necessary, clean the electrode following the manufacturer's instructions.

2. **Calibrate the pH meter**: Calibrate the pH meter **using standard buffer solutions**. Follow the instructions provided by the manufacturer to perform the calibration accurately. Typically, a two-point calibration using **pH 7** and **pH 4** or **pH 10** buffers is recommended.

3. **Rinse the electrode**: Rinse the electrode **with distilled or deionized water** to remove any residual buffer solution from the previous calibration. Gently blot the electrode **dry with a clean tissue or lint-free cloth**.

4. **Immerse the electrode**: Immerse the electrode into the solution being measured. Ensure that the electrode is fully submerged and that the **liquid level reaches the designated immersion line** on the electrode.

5. **Allow stabilization**: Allow the pH reading to stabilize. This typically takes a few seconds or until the pH reading becomes steady. **Avoid stirring the solution vigorously during stabilization**.

6. **Take the measurement**: Record the pH reading displayed on the pH meter. Ensure accurate readings by reading the value at eye level and **avoiding air bubbles on the electrode surface**.

7. **Rinse and store the electrode**: After each measurement, rinse the electrode **with distilled or deionized water** to remove any residues. Gently blot the electrode dry and store it in a proper storage solution or as recommended by the manufacturer.

8. **Maintain electrode integrity**: Regularly **clean** and **maintain** the electrode according to the manufacturer's guidelines. **Periodically check** for signs of wear or damage, and replace the electrode when necessary.

9. **Store the pH meter properly**: When not in use, store the pH meter with its electrodes submerged in a **3 M potassium chloride (KCl) solution or a pH 4 buffer solution**. This helps **prevent drying out** and **maintains electrode performance**. Avoid storing the electrodes in distilled water, as it can cause leaching or damage.

Troubleshooting

In certain scenarios, the operation of a pH meter may be subject to certain errors. Among these, two commonly encountered errors are the **sodium error** and **temperature error**, which will be elaborated upon in the following discussion.

Sodium error

The phenomenon known as sodium error can result in a **decrease in pH measurement by approximately 0.4-0.5 units**. It occurs particularly at **highly basic or alkaline pH levels**, where the concentration of **hydrogen ions (H+) is relatively low** compared to the concentration of sodium ions (Na+) present in the sample.

Causes

Sodium error arises **when the level of sodium ions is significantly high**, leading to the **replacement of some H+ ions** in the gel layer surrounding the sensitive electrode membrane with Na+ ions. Consequently, the **electrode's** response becomes influenced by sodium ions rather than hydrogen ions, resulting in a falsely lower pH reading than the actual pH value.

Prevention

To mitigate sodium error, it is advisable to consider the specifications of your pH electrode. Most **electrodes indicate the pH range they can accurately measure** and may also specify the **pH value at which sodium error begins** to occur, **typically around pH 12 or higher**. Whenever feasible, it is recommended **to measure highly alkaline solutions at room temperature**, as elevated temperatures can exacerbate the alkaline or sodium error effect.

Temperature effect

The pH of a buffer solution can be influenced by changes in temperature. This occurs **because the dissociation constant (pKa) of ions in the solution is temperature-dependent**. For instance, in the case of Tris Buffer, there is a change in pH with temperature at a rate of -0.031 pKa/°C. This means that if a pH 7.0 Tris Buffer is prepared at 40°C, its pH will be 5.95 at 37°C.

Causes

The temperature effect on pH arises due to the inherent nature of ions in the solution. The **dissociation constant (pKa) of these ions changes with varying temperatures**. As a result, the pH of the buffer solution experiences fluctuations when subjected to different temperature conditions.

Prevention

To mitigate the temperature effect on pH measurements, **it is essential to prepare the buffer solution at the same temperature at which it will be used**. Additionally, it is advisable to **calibrate the pH electrode** using buffers that are at the same temperature as the solution being measured. This ensures accurate pH readings by **minimizing the impact of temperature-induced changes in the buffer's pH**.

Note

In a simple definition, the dissociation constant indicates how easily a substance dissociates into ions when placed in a solution. A higher dissociation constant value suggests a greater degree of dissociation, while a lower value indicates less dissociation.

Weighing balance

A weighing balance is a scientific instrument utilized to **determine the weight or mass of an object**. It plays a vital role in various settings such as laboratories, commercial establishments, kitchens, and pharmacies. Weighing balances are available in a diverse range of sizes, each with its specific weighing capacity.

Fundamental concept

The working principle of a weighing balance involves the **utilization of an electromagnetic coil and a magnetic field created** by an amplifier. The **weighing pan**, to which the object is placed, **is connected to the electromagnetic coil**. This **coil is responsible for generating an electric current** that flows through it.

The magnetic field created by the amplifier interacts with the electromagnetic coil. The **aim is to maintain a state of balance between the lever** (remember that the balance operates based on the lever principle) and **the mass of the weighing pan.** As **additional weight is applied to the pan**, the **electric current passing through the coil is increased** to maintain the position of the lever.

The **counteracting force generated by this process is carefully measured** and **translated** by various electronic components within the weighing balance. These **components work together to produce a readable result**. The resulting electrical current is then further processed and displayed as a numerical value that is shown to the user.

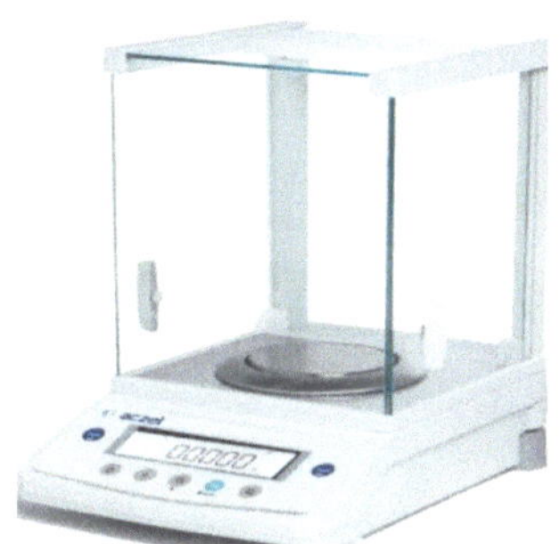

Protocol

1. **Ensure a stable environment**: Place the weighing balance on a **stable** and **level surface**, away from sources of vibration, drafts, or air currents that could affect the accuracy of the measurements.

2. **Zeroing the balance**: Before starting any measurements, make sure to **zero the balance**. Press the **"tare"** or **"zero"** button to **eliminate any residual weight** on the weighing pan. This ensures that subsequent measurements are based solely on the weight of the sample.

3. **Handling samples**: When placing samples on the weighing pan, **use appropriate containers** or **weighing boats** to prevent direct contact between the sample and the balance. This helps maintain the cleanliness of the balance and prevents cross-contamination.

4. **Avoid overloading**: Ensure that the **weight of the sample being measured is within the specified weighing capacity** of the balance. Overloading the balance can result in inaccurate readings or damage to the equipment.

5. **Gentle handling**: Handle samples and weigh containers with care to avoid spills or loss of material. Use **clean, dry forceps or spatulas to transfer samples**, minimizing direct contact with the balance.

6. **Allow for stabilization: After placing a sample** on the weighing pan, give it a few moments to stabilize. This **allows any residual air currents or vibrations caused by the sample placement to settle**, ensuring accurate measurements.

7. **Record measurements accurately**: Once the balance has stabilized and provided a stable reading, **record the weight promptly**. Avoid prolonged exposure of the sample on the weighing pan, as environmental factors or sample evaporation can lead to weight fluctuations.

8. **Clean and maintain regularly**: Keep the weighing balance clean by using appropriate cleaning agents and methods recommended by the manufacturer. Regular maintenance, such as **calibration checks** and **servicing**, should be performed as per the **manufacturer's guidelines** to ensure accurate and reliable measurements.

9. **Follow laboratory guidelines**: Adhere to any specific guidelines or procedures set by the laboratory regarding the use and maintenance of weighing balances. This ensures **consistency** and **compliance** with established protocols.

Laminar airflow

Laminar air flow refers to a **controlled and uniform airflow in a specific direction within a confined space**, such as a laboratory. It is achieved by passing the air through a **High-Efficiency Particulate Air Filter (HEPA)** that removes particles, ensuring a clean and sterile environment. The laminar airflow maintains a **steady and unidirectional stream, minimizing turbulence** and **preventing the introduction of contaminants**.

Significance of laminar airflow in the laboratory

Laminar air flow systems play a crucial role in various laboratory settings, including microbiology, pharmaceuticals, electronics, and research laboratories. The significance of laminar air flow can be summarized as follows:

1. **Contamination control**: Laminar air flow creates a **sterile working environment** by continuously supplying filtered air in a controlled manner. This helps to **prevent the entry of airborne particles, microorganisms**, and **contaminants**, thus **minimizing the risk of cross-contamination** and maintaining the integrity of sensitive experiments or processes.

2. **Protection for samples and personnel**: Laminar air flow protects both laboratory personnel and samples. By creating a clean air environment, it **safeguards valuable samples, prevents contamination of cultures**, and **maintains the sterility of equipment and surfaces**. It also helps protect researchers by **reducing exposure to hazardous substances or airborne pathogens**.

3. **Maintenance of product quality**: In industries such as pharmaceuticals and electronics, laminar airflow is essential for maintaining the quality and integrity of products. It **prevents the deposition of dust, particulates**, and **microorganisms** during manufacturing processes, ensuring product consistency and minimizing defects.

Protocol

1. **Preparing the work area**: Clean the work surface thoroughly before starting any procedures. **Remove any unnecessary items** and **disinfect the surface** using appropriate cleaning agents. Make sure the **laminar flow hood is powered on** and the **filters are clean and functioning properly**.

2. **Personal protective equipment (PPE)**: Wear appropriate PPE, including **gloves**, a **lab coat**, and a **face mask** or **respiratory protection**, as required by the laboratory's safety guidelines.

3. **Sterilize work tools**: Sterilize all necessary tools, such as **forceps, pipettes, or scalpels**, using appropriate methods such as **autoclaving or chemical disinfection.**

4. **Hand hygiene**: Wash hands thoroughly with **soap and water** or use an appropriate **hand sanitizer** before working within the laminar flow hood.

5. **Sample preparation**: Prepare your samples in a separate, clean area outside the laminar flow hood. Ensure they are **properly sealed** and **free from any contaminants**.

6. **Working within the laminar flow hood**: Place your **hands inside the hood, maintaining a distance from the filter face**, to **minimize disruption of the laminar airflow. Avoid unnecessary movements** and **minimize talking** to reduce the generation of particles.

7. **Sterile techniques**: Follow sterile techniques while handling samples or conducting procedures within the laminar flow hood. This includes using **aseptic techniques (Spray 75% ethanol at the workplace)**, **flame sterilization**, and **avoiding cross-contamination**.

8. **Clean as you go**: Regularly clean the work surface and equipment inside the laminar flow hood using **appropriate disinfectants** to maintain a sterile environment.

9. **Proper waste disposal**: Dispose of all waste materials, including used **samples, contaminated materials**, and **consumables**, in designated waste containers following laboratory protocols.

10. **Shutdown procedure**: After completing the work, **power off the laminar flow hood, clean the work surface again**, and perform any necessary maintenance or filter replacements as per the manufacturer's instructions.

Water distillation unit

A water distillation unit is a laboratory instrument used to purify water through the process of **distillation**. It serves the crucial purpose of **obtaining highly pure water, free from impurities** and **contaminants**. This unit is particularly significant in laboratory settings where the quality of water plays a critical role in experiments, analyses, and various scientific procedures.

Significance of the water distillation unit in the laboratory

In laboratory settings, water serves as a fundamental component for a wide range of applications, including **solution preparation**, **sample dilution**, **media preparation**, and **equipment cleaning**. The quality of water used in these processes can significantly impact the accuracy, reliability, and reproducibility on experimental results.

Water distillation units are specifically designed to produce highly pure water by **removing various impurities** such as minerals, organic compounds, bacteria, and other contaminants. This ensures that the water used in laboratory procedures meets the required standards and is free from substances that may interfere with experiments or compromise the validity of results.

Protocol

1. **Preparation**: Ensure that the **water distillation unit is clean** and free from any residue or contaminants. **Check the unit for any signs of damage** or leaks before proceeding.

2. **Water Source**: Connect the water distillation unit to a reliable water source, such as a **water tap or a purified water supply**. Ensure that the water source is clean and free from contaminants.

3. **Fill the Boiler**: Fill the boiler of the distillation unit with **an appropriate volume of water**, taking care not to exceed the maximum capacity specified by the manufacturer. **Close the boiler securely**.

4. **Start the Distillation Process**: **Switch on** the distillation unit and **set the desired temperature or distillation mode**, following the manufacturer's instructions. The water will begin to heat up, and the distillation process will commence.

5. **Collection**: As the distillation process progresses, the water vapor will condense and **collect in a separate container or receiving flask**. **Ensure that the receiving flask is clean** and suitable for storing distilled water.

6. **Monitoring**: Monitor the distillation process closely, **keeping an eye on the water level** in the boiler and the progress of water collection in the receiving flask. Make sure there are no interruptions or issues during the distillation process.

7. **Completion**: Once the desired volume of distilled water has been collected, **turn off the distillation unit and allow it to cool down** before handling or cleaning. **Safely disconnect the unit** from the water source.

Note

Do not get confused with the below terms!

1. Single Distilled Water: Single distilled water refers to water that has undergone one cycle of distillation, removing a significant portion of impurities. It is suitable for general laboratory applications but may still contain traces of certain impurities.

2. Double Distilled Water: Double distilled water refers to water that has undergone two cycles of distillation, further purifying it and reducing impurities to an even greater extent. It is typically used for more sensitive laboratory procedures.

3. Nuclease-Free Water: Nuclease-free water is water that has been treated to ensure the absence of nucleases and enzymes that can degrade DNA and RNA. It is commonly used in molecular biology and genetics research to minimize the risk of nucleic acid degradation.

Centrifuge

The centrifuge process is a technique used in laboratories **to separate substances of different densities or particle sizes by spinning them at high speeds**. It relies on the principle of centrifugal force, which causes denser particles to move towards the outer edges of a rotating container, while lighter particles remain closer to the center.

Types of centrifuges

1. **Microcentrifuge**: This type of centrifuge is used **for small-volume samples** and can reach high speeds, typically generating forces of **up to 20,000 to 30,000 times the force of gravity (20,000-30,000 g)**. It is commonly employed for **quick spins, DNA/RNA extraction**, and **protein precipitation**.

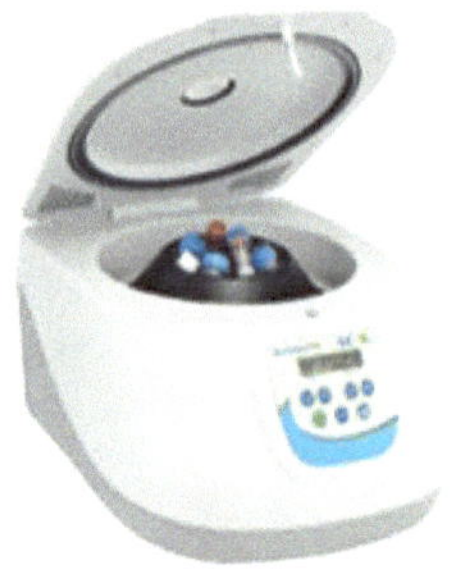

2. **Benchtop Centrifuge**: It is a versatile centrifuge suitable for a **wide range of laboratory applications.** It can accommodate various rotor types and offers a balance between speed and capacity, typically generating forces of **up to 10,000 to 20,000 times the force of gravity (10,000-20,000 g).**

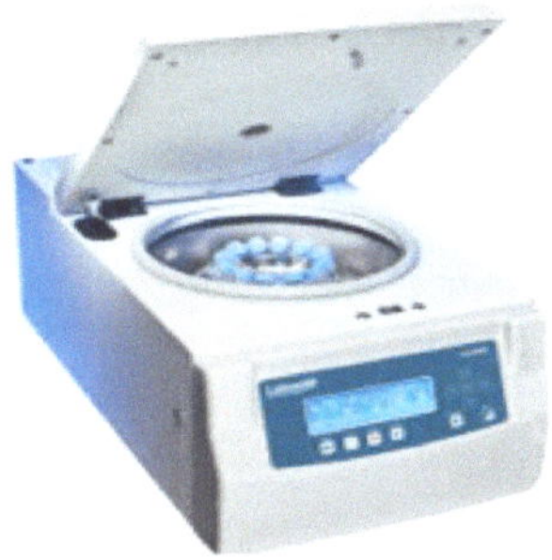

3. **Refrigerated Centrifuge**: Designed for **temperature-sensitive samples,** this centrifuge includes **a cooling system to maintain low temperatures** during centrifugation. It is crucial for preserving sample integrity in applications like **cell culture** and **enzyme assays**. It can generate forces of **up to 20,000 to 25,000 times the force of gravity (20,000-25,000 g).**

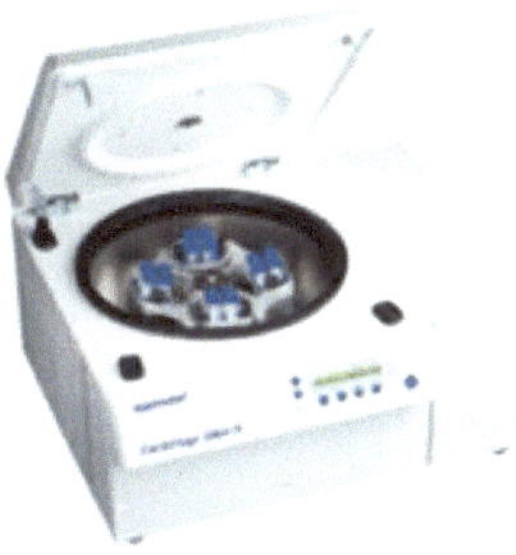

4. **Ultracentrifuge**: Ultracentrifuges operate at **extremely high speeds** and are used for applications such as **density gradient centrifugation, isolation subcellular components,** and **separation of macromolecules** based on size and shape. They can generate forces ranging from **100,000 to 1,000,000 times the force of gravity (100,000-1,000,000 g) or even higher.**

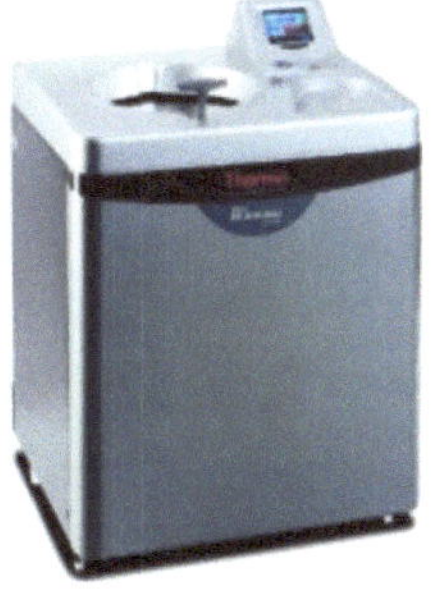

5. **Floor-standing Centrifuge**: This centrifuge is **larger and more powerful**, with **high capacities** and **capabilities for heavy-duty** applications in research, biotechnology, and industrial laboratories. It can generate forces ranging from **10,000 to 50,000 times the force of gravity (10,000-50,000 g) or higher**, depending on the specific model.

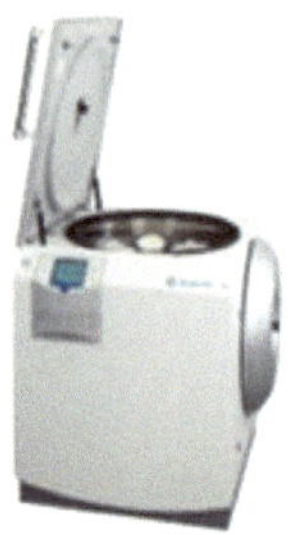

Fundamental concept

When a particle subjected to centrifugation, it experiences mainly **three types of forces** that is **Buoyancy**, **Frictional force**, and **Centrifugal force**. When the centrifugal force is more than the sum of frictional force and buoyancy the particle experience sedimentation, whereas if the centrifugal force is equal to the sum of frictional force and Buoyancy it experience separation based on density gradient.

If,

F_1 = Centrifugal Force

F_2 = Frictional Force

F_3 = Buoyancy

Then,

Sedimentation: $F_1 > F_2 + F_3$

Density Gradient: $F_1 = F_2 + F_3$

Significance

The significance of centrifugation in the laboratory lies in its ability to separate different components of a sample, such as cells, organelles, proteins, or nucleic acids, based on their sedimentation rates. This technique allows for the isolation and purification of specific components, aiding in various research, diagnostic, and industrial applications.

Protocol

1. **Preparing the centrifuge**: Ensure the centrifuge is clean and free from any residues that could contaminate the samples. Check that the rotor is securely fastened to the centrifuge and properly balanced. Verify that the centrifuge is plugged in and connected to a power source.

2. **Loading the samples**: Select appropriate centrifuge tubes or containers suitable for your samples. Label the tubes with the necessary information for identification. Distribute the samples evenly among the tubes, ensuring the tubes are balanced. Ensure the lids or caps of the tubes are tightly closed to prevent sample leakage.

3. **Balancing the centrifuge**: Place the tubes symmetrically in the rotor to achieve balance. If using multiple rotor cavities, distribute the tubes equally across them. Use counterweights or additional empty tubes to balance the rotor, if required.

4. **Setting the parameters**: Consult the user manual or the specific protocol for recommended centrifugation settings such as speed, time, and temperature. Set the desired centrifugation parameters on the control panel of the centrifuge.

5. **Starting the centrifuge**: Close the centrifuge lid or cover securely. Double-check that the rotor is properly inserted and secured. Start the centrifuge according to the manufacturer's instructions.

6. **After centrifugation**: Once the run is complete, wait for the centrifuge to come to a complete stop before opening the lid. Carefully remove the tubes from the rotor, taking precautions to avoid sample cross-contamination. Handle the tubes gently to prevent disruption of any pellet formed during centrifugation. Proceed with further sample processing or analysis as required.

Precaution

To ensure the safety of both the centrifuge instrument and personnel, it is crucial to **properly balance the sample before starting the centrifuge**. The balancing process may vary depending on the centrifugal force applied during centrifugation. Here's a detailed explanation of each step:

1. **Centrifugation up to 5,000g**: For centrifugation at lower forces, you can achieve balance **by pouring an equal volume of liquid** into other centrifuge tubes. This method ensures that the weight distribution is even across the tubes, minimizing the risk of imbalance. By maintaining balanced samples, you prevent excessive vibrations and potential damage to the centrifuge or injury to personnel.

2. **Centrifugation beyond 12,000g**: When operating at higher centrifugal forces, it is important to achieve a more precise balance. To accomplish this, you should **use a weighing balance with milligram accuracy**. First, weigh each sample carefully on the balance, ensuring their weights are as close to each other as possible. Then, arrange the samples symmetrically within the centrifuge rotor to distribute the weight evenly. This meticulous balancing helps mitigate the effects of centrifugal forces and reduces the likelihood of imbalances that could lead to instrument malfunction or accidents.

3. **Centrifugation at ultra-speed**: When working with ultra-speed centrifuges, **balancing becomes even more critical**. These centrifuges operate at very high speeds, generating significant forces. Therefore, it is **crucial to achieve precise and accurate balancing** to minimize any vibrations or deviations that could compromise the instrument's performance or pose risks to safety. Follow the manufacturer's guidelines closely and take extra care in achieving perfect balance, as even minor imbalances can have pronounced effects at ultra-speeds.

4. **Centrifugation at very low temperatures (e.g., 4°C)**: When using a centrifuge at low temperatures, such as in refrigerated centrifuges, **condensation can occur on the inner surfaces of the centrifuge cups or tubes after centrifugation**. To avoid potential damage and ensure accurate subsequent experiments, it is essential to **wipe the centrifuge cup or tube with a dry cloth to remove any moisture**. Additionally, **leaving the lid open** allows the **remaining water droplets to evaporate fully** before storing the instrument. This practice helps maintain the integrity of the centrifuge and prevents any water-related issues during future use.

Cleaning laboratory glassware

Properly cleaning laboratory glassware is essential for maintaining accurate and reliable experimental results. The importance of cleaning glassware can be understood through the following points:

Importance of cleaning laboratory glassware

1. **Chemical and Biochemical Contamination**: Many chemicals and biochemicals used in laboratory experiments are utilized in milligrams or microgram amounts. Any contamination present in the glassware can **significantly affect the total experimental sample**, potentially leading to inaccurate results. It is crucial to prevent the introduction of contaminants during the experimental process.

2. **Sensitivity to Contaminants**: Numerous biochemical and biochemical processes are **sensitive to common contaminants, including metal ions, detergents**, and **organic residues**. These contaminants can interfere with the desired reactions or introduce unwanted reactions, compromising the accuracy and reliability of the experimental outcomes.

Prevention

To ensure proper cleaning and prevention of contamination, the following practices can be followed:

1. **Removal of Organic and Metal Ion Contamination**: Wash glassware with a **dilute detergent solution (0.5% in water)**, followed by **several rinses with distilled or deionized water**. This procedure is generally sufficient for routine cleaning purposes. **To reduce metal ion contamination**, rinse glassware with **concentrated nitric acid**, followed by **extensive rinsing with purified water**.

2. **Cleaning Optically Polished Glassware**: Avoid cleaning cuvettes or any optically polished glassware **with strong bases like ethanolic KOH**, as it can cause **etching**. Instead, carefully clean cuvettes with a **0.5% detergent solution, using a sonicated bath or cuvette washer**.

3. **Special Procedures for Glass Pipets**: Immediately after use, **place pipets vertically, tip-up, in a vertical cylinder containing a dilute detergent solution (less than 0.5%)**. Ensure that the entire pipet is covered with the solution. Transfer the pipets to a separate container after several have accumulated in the detergent solution. For glass **pipets contaminated with proteins, chromic acid treatment followed by thorough rinsing with pure water** can be effective.

Preparation of water

Water is a fundamental component in various laboratory procedures, and its quality directly impacts experimental results. Purifying water is essential **to eliminate impurities** that can interfere with experiments or compromise the accuracy of measurements. Here is a detailed explanation of the reasons behind the need for water purification:

1. **Impurities in Ordinary Tap Water**: Ordinary tap water typically contains a range of impurities, including **particulate matter** such as **sand** and **silt**. Additionally, it may contain dissolved organic and inorganic substances, as well as microorganisms. These impurities can introduce contaminants into experiments, leading to inaccurate results or interfering with the intended processes.

2. **Metabolic By-Products from Microorganisms**: Tap water may also contain metabolic by-products from microorganisms, known as **pyrogens**. **Pyrogens are substances that can cause fever when introduced into the body**. In laboratory settings, the presence of pyrogens can **affect the reliability and safety of experiments**, making water purification crucial.

Methods of water purification

Several methods are employed to purify water in laboratories. The choiced method depends on the specific requirements and applications. Here are some commonly used water purification methods:

1. **Distillation**: Distillation involves **boiling the water** and **then condensing the vapor to obtain purified water**. This process effectively removes most impurities, including dissolved solids, microorganisms, and pyrogens. Distillation is a reliable method for obtaining high-purity water in the laboratory.

2. **Ion Exchange**: Ion exchange involves **passing water through an ion exchange resin**, which selectively removes ions and replaces them with other ions of similar charge. This process can effectively **remove dissolved inorganic impurities from water**.

3. **Carbon Adsorption**: Carbon adsorption utilizes **activated carbon filters** to **remove organic impurities** from water. The activated carbon has a large surface area, allowing it to adsorb organic molecules. This method is effective in removing dissolved organic compounds and improving water quality.

4. **Reverse Osmosis**: Reverse osmosis employs a **semipermeable membrane** to remove impurities from water. It works **by applying pressure to force water through the membrane, leaving behind most dissolved solids,**

microorganisms, and **pyrogens**. Reverse osmosis is particularly effective in removing contaminants at the molecular level.

5. **Membrane Filtration**: Membrane filtration involves passing water through a **fine membrane** that acts as a physical barrier to remove particles, microorganisms, and some dissolved impurities. This method can effectively purify water by **retaining impurities on the membrane surface** and allowing purified water to pass through.

Chapter 3

Solution And Buffer Preparation

Introduction

Understanding the proper handling and operation of laboratory instruments, which are discussed in the second chapter lays a solid foundation for the subsequent steps involved in preparing solutions and buffers. By mastering the techniques of instrument operation, you will be equipped with the necessary skills to accurately measure and manipulate the various components required for solution preparation.

The solution is a homogeneous liquid blend comprising a primary component known as the solvent, which forms most of the mixture, and a secondary component referred to as the solute, which is evenly dispersed throughout the solvent. This molecular arrangement ensures that the solute particles are uniformly distributed within the solvent, resulting in a well-mixed solution. Some of the most common units used for preparing the solution are given below:

1. **Molarity (M)**: Molarity represents the **concentration of a solution in terms of moles of solute per liter of solution**. For instance, if you are tasked with preparing a 1M glucose solution in a 100 ml volume, you would need to weigh 18g of glucose (considering its molecular weight as 180) and transfer it to a 1-liter volumetric flask.

2. **Normality (N)**: Normality refers to the **concentration of a solution expressed as the number of equivalents of solute per liter of solution**. This unit is particularly relevant **when dealing with acid-base reactions or redox reactions**, where the concept of equivalents is essential for stoichiometric calculations. For example, a 0.5N sulfuric acid (H_2SO_4) solution contains half the number of equivalents of sulfuric acid per liter compared to its molar concentration.

3. **Percent by weight (%wt/wt)**: Percent by weight denotes the **concentration of a solution based on the mass of solute per 100 grams** of the total solution mass. Suppose you want to prepare a 5%wt/wt sodium chloride (NaCl) solution. This means you would dissolve 5 grams of NaCl in 100 grams of the total solution mass, resulting in a final solution with a total mass of 105 grams.

4. **Percent by volume (%wt/vol)**: Percent by volume represents the **concentration of a solution in terms of the mass of solute per 100 milliliters of the solution volume**. As an example, to create a 2%wt/vol hydrogen peroxide (H_2O_2) solution, you would mix 2 grams of hydrogen peroxide with enough water to make a total volume of 100 ml.

5. **Weight per volume (wt/vol)**: Weight per volume describes the **concentration of a solution by specifying the mass of solute (in grams, milligrams, or micrograms) per unit volume (e.g., mg/ml, g/liter, mg/100ml)**. For instance, a 50 mg/ml penicillin solution implies that there are 50 milligrams of penicillin dissolved in each milliliter of the solution.

Virtual learning aid

Note

Molarity is the concentration unit that is frequently used in laboratory experiments. It is important to have a clear understanding of this concept as it plays a crucial role in successfully conducting a wide range of practical experiments in the lab.

pH

pH (Potential of Hydrogen) is a fundamental concept that **measures the acidity or alkalinity of a solution**. It is defined as the **negative logarithm (base 10) of the hydrogen ion concentration in a solution**. The **pH scale ranges from 0 to 14**, where values **below 7 indicate acidity**, values **above 7 indicate alkalinity** and a **pH of 7 represents neutrality**. To better understand the concept, let's take an example. Imagine a solution with a pH of 3. This means the hydrogen ion concentration in the solution is 10^{-3} moles per liter. As the pH scale is logarithmic, each unit change in pH represents a tenfold difference in hydrogen ion concentration. Therefore, a solution with a pH of 2 will have ten times higher hydrogen ion concentration compared to a pH of 3, and a solution with a pH of 1 will have one hundred times higher concentration.

pH has significant implications in various scientific fields as it affects chemical reactions, enzyme activity, nutrient availability, and even the survival of living organisms. In laboratory settings, understanding pH is crucial for adjusting and controlling the acidity or alkalinity of solutions, conducting experiments, and interpreting experimental results accurately.

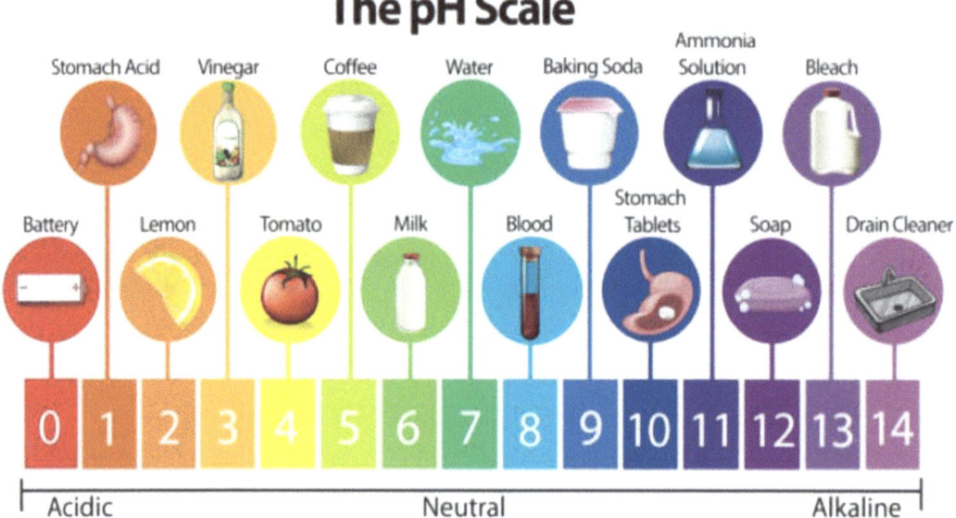

BUFFER

Buffer solutions play a crucial role in **maintaining a stable pH in various chemical applications**, serving as effective regulators. In natural systems, such as the bicarbonate buffering system in the blood, a delicate balance is maintained by the presence of Carbonic acid (H_2CO_3), Bicarbonate ion (HCO_3^-), and Carbon dioxide (CO_2). This equilibrium ensures the blood pH remains within the optimal range of 7.35 to 7.40, supporting essential metabolic functions.

Why there is a need for Buffer?

Buffers are essential in controlling biological processes due to their role in maintaining pH levels. For instance, enzymes, which are composed with ionizable amino acids, rely on specific ionization states for substrate binding, catalytic intermediates formation, and product release. Examples include pepsin, with an optimum pH of 1.5 in the stomach, and trypsin, which functions best at pH 7.4. In pathological conditions like diabetes or fasting, excessive production of β-hydroxybutyric acid from fat metabolism can lower blood pH, leading to enzyme dysfunction and symptoms such as headaches, nausea, and convulsions. Additionally, bicarbonate buffer systems are crucial for regulating blood pH and play a vital role in respiration.

How does buffer maintain pH?

A buffer solution consists of a combination of a weak acid and its conjugate base or a weak base and its conjugate acid. This equilibrium system is represented by the reversible ionization equation:

$HA + B^- \rightleftharpoons A^- + HB$.

Here, HA and A^- form one conjugate acid-base pair, while B^- and HB form another pair. When a strong acid is added, it introduces additional H^+ ions. In response, the equilibrium shifts to the left, resulting in the formation of more undissociated HA and minimizing the change in H^+ concentration. Conversely, when a strong base, such as OH^-, is added, it consumes H^+ ions. This causes the equilibrium to shift to the right, producing more A^- ions and minimizing the change in pH.

The buffer capacity quantitatively measures the ability of a buffer solution to resist changes in pH when an acid or alkali is introduced. It indicates the solution's resistance to shifts in acid or base concentration, ensuring that the pH remains relatively stable.

Henderson-Hasselbalch equation

The dissociation of a weak acid in water can be written as a reversible reaction:

$HA + H2O \rightleftharpoons H3O^+ + A^-$

The ionization can be represented by the equilibrium constant, K_{eq} :

$$K_{eq} = [H_3O^+] \, [A^-] / [H_2O] \, [HA]$$
$$pH = pK_a + \log[A^-/AH]$$

The above equation is also known as the **Henderson-Hesselbalch equation**. From this equation, **pK_a can be defined as the pH at which an acid is 50% ionized**.

Note

Example

How to prepare 200 mM Phosphate Buffer (pH 7.4) using Phosphoric Acid and Sodium Dihydrogen Phosphate?

To prepare 500 ml of a 200 mM phosphate buffer with a pH of 7.4, we will use the chemicals phosphoric acid (H_3PO_4) and sodium dihydrogen phosphate (NaH_2PO_4). The procedure involves utilizing the Henderson-Hasselbalch equation, which allows us to select the appropriate ratio of these chemicals to achieve the desired pH.

Step 1: Understanding Phosphoric Acid and Its Dissociation

Phosphoric acid (H_3PO_4) is a triprotic acid, meaning it can release three protons successively in a stepwise manner. These steps involve the dissociation of H3PO4 into its conjugate bases: dihydrogen phosphate ($H_2PO_4^-$) hydrogen phosphate (HPO_4^{2-}), and phosphate (PO_4^{3-}).

The dissociation constants (Ka values) for phosphoric acid are as follows:

Ka1 = 7.6×10^{-3}

Ka2 = 6.16×10^{-8}

Ka3 = 4.79×10^{-13}

Step 2: Selecting the Appropriate pKa Value for Buffer Preparation

To create a phosphate buffer with a pH of 7.4, we **choose the pKa value that aligns with the desired pH**. In this case, pKa2 (pKa value of the second dissociation step) is closest to 7.4, making it suitable for our buffer.

Step 3: Calculating the Ratio of Conjugate Base to Acid

Using the Henderson-Hasselbalch equation, we can determine the ratio of the concentrations of hydrogen phosphate (HPO_4^{2-}) to dihydrogen phosphate ($H_2PO_4^-$) required for our buffer.

pH = pKa + log([HPO$_4^{2-}$]/[H$_2$PO$_4^-$])
7.4 = 7.21 + log([HPO$_4^{2-}$]/[H$_2$PO$_4^-$])
log([HPO$_4^{2-}$]/[H$_2$PO$_4^-$]) = 0.19
[HPO$_4^{2-}$]/[H$_2$PO$_4^-$] = $10^{0.19}$ ≈ 1.5488
Thus, the ratio of hydrogen phosphate to dihydrogen phosphate needed for the buffer is approximately 1.55:1.

Step 4: Calculating the Amount of Chemicals Required

We want to prepare 500 ml of a 200 mM buffer. To calculate the amount of each chemical required, we first find the number of moles of the buffer components.
Number of moles required for the 500 ml buffer = Molarity of buffer × Volume (in liters)
Number of moles required = 0.2 M × 0.5 L = 0.1 moles
Now, we calculate the moles of each component based on the previously determined ratio:
Moles of NaH$_2$PO$_4$ required = (1/1.55) × 0.1 ≈ 0.0645 moles
Moles of Na$_2$HPO$_4$.7H$_2$O required = (1.55/1.55+1) × 0.1 ≈ 0.0355 moles

Step 5: Determining the Mass of Chemicals Needed

Finally, we convert the moles to grams using the molecular weights of the chemicals.
Mass of Na$_2$HPO$_4$ required = 0.0645 moles × 119.98 g/mol ≈ 7.74 grams
Mass of Na$_2$HPO$_4$.7H$_2$O required = 0.0355 moles × 268.07 g/mol ≈ 9.52 grams
    Therefore, to prepare 500 ml of 200 mM phosphate buffer at pH 7.4, we need approximately 7.74 grams of sodium dihydrogen phosphate (Na$_2$HPO$_4$) and 9.52 grams of sodium hydrogen phosphate heptahydrate (Na$_2$HPO$_4$.7H$_2$O). By following this procedure, you can confidently prepare the desired buffer for your experiments.
1. Carefully weigh 7.74 grams of Na$_2$HPO$_4$ and 9.52 grams of Na$_2$HPO$_4$.7H$_2$O using a weighing balance.
2. Transfer the weighed salts into a 1-liter conical flask.
3. Add 400 ml of distilled water to the conical flask containing the salts.
4. Shake the flask vigorously to ensure complete dissolution of the salts in the water.
5. Thoroughly wash the pH meter electrode with distilled water to ensure accurate pH measurement.
6. Place the clean pH electrode into the phosphate solution and measure the pH of the solution.
7. While we aim for a pH of 7.4, the initial pH of the solution may differ.
8. If the measured pH is below 7.4, add 1 N NaOH dropwise to the solution, shake the flask well, and measure the pH again. Repeat this process until a pH of 7.4 is achieved.

9. If the measured pH is above 7.4, add 1 N H3PO4 dropwise to the solution, shake the flask well, and measure the pH again. Continue this process until a pH of 7.4 is reached.

10. Once the desired pH of 7.4 is achieved, the buffer solution is ready to be used in your experiments!

Chapter 4

Electrophoresis

Introduction

In our quest to uncover the mysteries of biotechnology, we have gained a solid understanding of essential concepts such as Good Laboratory Practice, buffers, and pH in the previous chapters. Now, we embark on an exciting journey into the world of a routinely used laboratory technique known as electrophoresis. As we continue this exploration together, we will unravel the secrets of various biotechnological tools and procedures.

Electrophoresis derives its name from **two Greek words**: "Electro," representing the energy of electricity, and "**Phoresis**," which translates to "carry across." In simple terms, it refers to **a method for separating charged molecules in a mixture by applying an electric field**. This powerful tool is widely used in life sciences to analyze and **resolve proteins**, **DNA**, and other **charged biomolecules**. When subjected to an electric field, charged molecules move or **migrate at a speed dictated by their charge-to-mass ratio**. This migration allows scientists to separate the different biomolecules present in a sample based on their size and charge characteristics. In practical terms, electrophoresis finds significant applications in life sciences research.

Mathematical aspect

According to the laws of electrostatics, an ion with **charge 'Q'** in an electric field of strength 'E' will experience an **electric force, $F_{electrical}$**

$$F_{electrical} = Q.E$$

The resulting migration of charged molecule through the solution is opposed by a **frictional force, $F_{frictional}$**

$$F_{frictional} = V. f$$

Where **V is the rate of migration of charged molecule** and f **is the frictional coefficient**.

The **frictional coefficient** measures the resistance encountered by the molecule in moving through the solvent. It **depends on the size** and **shape of the**

migrating molecules and the **viscosity of the medium**. For a spherical molecule, it is given by **Stroke's law**:

$$f = 6\,\pi\,\eta\,r$$

Where η **is the viscosity** of the solvent and **r is the radius** of the molecule. In the constant electric field, the force on charged molecules balances each other;

$$Q.E = V.\,f$$

So that each charged molecule moves with a constant characteristic velocity. The migration of a charged molecule in the electric field is generally expressed in terms of **electrophoretic mobility (μ)**, which is the ratio of the migration rate of the charged molecule to the applied electric field:

$$\mu = V/E = Q/f$$

So according to the above equation, electrophoretic mobility is directly proportional to charge and inversely proportional to the viscosity of the medium, size, and shape of the molecule.

Electrophoresis comes in two types.

Moving boundary electrophoresis

Moving boundary electrophoresis involves separating charged molecules in a free solution **without a supporting medium**. Developed **by Tiselius** in 1937, this **method utilizes a glass tube containing the sample dissolved in a buffer solution**, which acts as an electrolyte to maintain the desired pH. **Electrodes are connected to the tube**, and **when an electric potential is applied, charged molecules move toward one of the electrodes**. Since different charged molecules migrate at varying rates, distinct interfaces or boundaries are formed between the leading edge of each charged molecule and the remaining mixture. This **enables the separation of different charged molecules present in the initial mixture**.

Zone electrophoresis

In **zone electrophoresis**, the sample is confined within **an inert matrix**, such as filter paper soaked in a buffer (**paper electrophoresis**) or a gel (**gel electrophoresis**). This matrix is crucial because as electric current passes through the electrophoresis solution, it generates heat, leading to band diffusion and mixing without a supporting matrix. The choice of supporting matrix depends on the specific molecules to be separated and the desired basis for their separation. By using these supporting matrices, zone electrophoresis allows for precise and controlled separation of molecules, enabling researchers to analyze and study different components within a mixture effectively.

Note

Gel electrophoresis is a classic example of zone electrophoresis and is categorized into two types: horizontal gel electrophoresis and vertical gel electrophoresis.

1. Horizontal Gel Electrophoresis: In this system, electrophoresis is performed continuously on a horizontal gel. The sample is loaded onto the gel, and an electric current is applied, causing the charged molecules to move through the gel. This method is commonly used for separating DNA, RNA, and proteins based on their size and charge.

2. Vertical Gel Electrophoresis: In contrast, vertical gel electrophoresis involves a discontinuous approach. The gel slabs are positioned between the upper and lower tanks, and the buffer connects them. Researchers make various modifications to the running conditions to answer specific analytical questions. This technique is versatile and allows for more complex separation and analysis of biomolecules.

Note

Running of gel electrophoresis: In a vertical gel electrophoresis system, two types of gels, the stacking gel, and the resolving gel, are utilized. The process begins with the preparation of the resolving gel solution, which is poured into the gel cassette for polymerization. To maintain a smooth surface and prevent interference from oxygen during polymerization, a thin layer of organic solvent, such as butanol or isopropanol, is added on top of the resolving gel. Once the resolving gel has polymerized, the stacking gel is poured over it. A comb is then inserted into the gel to create wells in the stacking gel, facilitating the construction of different lanes for the samples. Before loading the samples, they are prepared in a loading dye containing SDS and β-mercaptoethanol to denature the samples and ensure accurate separation during electrophoresis. The presence of glycerol in the loading dye helps in the smooth loading of the samples into the wells.

Gel electrophoresis

When electrophoresis is performed in acrylamide or agarose gels, the **gel serves as a size-selective sieve during separation**. **Low-percentage gels** are used to **resolve large proteins**, and **high-percentage gels** are used to **resolve smaller proteins**. When a protein moves through a gel in response to the electric field, the gel's pore size allows the smaller protein to travel faster than the larger proteins.

Acrylamide is the material of choice for preparing electrophoretic gels to separate proteins. The mechanism of acrylamide polymerization involves the combination of **Ammonium Persulfate** and **Tetramethylethylenediamine (TEMED)**, which **produces oxygen-free radicals**. These **radicals initiate the polymerization process of acrylamide monomers**, leading to the formation of a linear polymer. The linear monomers are then **interconnected by cross-linking with bis-acrylamide monomers, creating a three-dimensional mesh with pores**. The size of these pores is controlled by the concentration of acrylamide and the amount of bis-acrylamide used in the gel.

Buffer system

During electrophoresis, the **buffer ions** are carried through gel just like sample ions: **negatively charged ions towards the anode** and **positively charged ones towards the cathode**.

These **three basic types of buffers** required

1. **Gel casting buffer** (to cast the gel)
2. **Sample buffer** (to prepare the sample)
3. **Running buffer** (To fill electrode reservoirs)

Electrophoresis can be conducted under two conditions.

Denaturing Electrophoresis

In denaturing electrophoresis, proteins are subjected to denaturing conditions using an **anionic detergent called sodium dodecyl Sulfate (SDS)**. SDS **causes the loss of the protein's 3D structure**, effectively denaturing it. Additionally, SDS **imparts a uniform negative charge to the proteins**, allowing them to migrate toward the positive electrode during electrophoresis. This method is particularly **useful for separating proteins based on their molecular weights**.

Nondenaturing Electrophoresis

Nondenaturing electrophoresis is performed under **native conditions**, maintaining the **native protein conformation**, **subunit interactions**, and **biological activity**. In this method, buffer systems are employed to preserve the protein's native structure. **Proteins are separated based on their charge-to-mass ratios**, with each protein migrating toward the positive electrode due to its inherent charge. Unlike denaturing electrophoresis, SDS is not present in the electrophoresis buffer, and the sample loading buffer does not contain SDS or reducing agents. Furthermore, the samples are not subjected to boiling, ensuring that their native structures remain intact.

The reason behind the accumulation of protein bands in a smaller zone is that the **pH of the stacking gel is 6.8** and at this pH, glycine is moving slowly in the front whereas Tris-HCl is moving fast. As a result, the sample gets sandwiched between glycine-Tris and gets stacked in the form of a thin band. Therefore, the relative mobility of ions in the stacking gel is chloride ions **(leading ion) > protein sample > glycinate ions (trailing ion)**. As the sample enters into the **resolving gel with pH 8.8**, the glycine is now charged, it moves fast, and now the sample runs as per their charge and mass ratio. The relative rate of movement of anions in the separating gel is **chloride ions > glycinate ions > protein sample**.

Now discontinuous **Polyacrylamide Gel Electrophoresis (PAGE)** can be native or denaturing. In **discontinuous native PAGE**, proteins are prepared in a nonreducing, non-denaturing sample buffer, and electrophoresis is also performed in the **absence of denaturing and reducing agents**. In **discontinuous denaturing PAGE**, electrophoresis is performed in the **presence of denaturing detergent sodium dodecyl sulfate (SDS), called SDS-PAGE**.

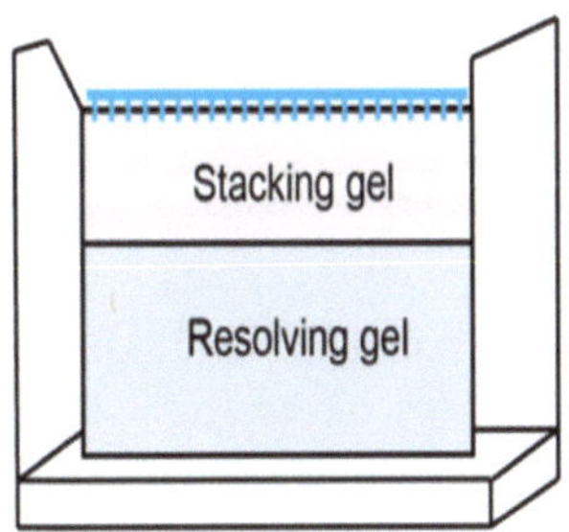

SDS-PAGE

SDS-PAGE is a widely used technique for separating and analyzing proteins and other biomolecules **based on their molecular weight**.

Fundamental Concept:

SDS-PAGE is a widely used technique in biochemistry to separate proteins based on their **charge density**, **mass**, and **shape**. To enable uniform migration through the gel, proteins are treated with the negatively charged detergent SDS, which imparts a large net negative charge on the proteins. This causes all SDS-bound proteins to move towards the positively charged electrode when an electric current is applied during gel electrophoresis. The migration rate is influenced by the protein's mass, with smaller proteins traveling faster due to the sieving effect of the gel matrix.

There are **two types** of SDS-PAGE: **reducing** and **non-reducing**. In reducing SDS-PAGE, along with SDS, proteins are treated with **reducing agents like β-mercaptoethanol to break disulfide bonds between cysteine residues**, ensuring the proteins are in their denatured, linear state. On the other hand, non-reducing SDS-PAGE does not use reducing agents, and as a result, the disulfide bonds within the proteins remain intact.

Virtual learning aid

Materials

1. Latex gloves
2. Weighing Balance
3. Graduated cylinder (200 ml)
4. Electrophoresis apparatus
5. Power pack
6. Pipettes
7. Eppendorf tubes
8. Boiling water bath
9. Magnetic stirring bar and stirrer
10. Beakers

Reagents

1. N, N, N', N'-Tetramethylethylenediamine (TEMED)
2. Ammonium Persulfate (APS)
3. Tris-HCl buffer
4. Glycine
5. Bromophenol Blue tracking dye
6. Coomassie Brilliant Blue R250
7. Sodium Dodecyl Sulfate (SDS)
8. Acrylamide
9. Bis-Acrylamide
10. β-mercaptoethanol
11. Glycerol
12. Ethanol/ butanol

Protocol

1. **Preparation of Running Buffer**: **Prepare a 1X Tris-HCl buffer** by diluting the stock solution with deionized water. **Ensure the pH is adjusted** to the desired value, usually around 8.8 for separating proteins. **Add an appropriate amount of Glycine** to the buffer solution (e.g., 0.1% w/v). Stir gently until dissolved. **Transfer the buffer to the electrophoresis apparatus**, making sure to fill both the upper and lower reservoirs.

2. **Gel Casting**: **Assemble the gel-casting apparatus**. For **a standard 10% resolving gel**, prepare a mixture of **Acrylamide** (e.g., 10% w/v), **Bis-Acrylamide** (e.g., 0.3% w/v), **1X Tris-HCl buffer**, **TEMED** (e.g., 0.1% v/v) - Use in a well-ventilated area due to its pungent odor, **APS** (e.g., 0.1% w/v). Add a small amount of **Bromophenol Blue tracking dye** to visualize the gel formation. Mix the components gently but thoroughly, **avoiding bubble**

formation. **Pour the gel solution into the gel-casting tray**, leaving space for the comb. **Insert a comb to create sample wells**. Allow the gel to polymerize **for about 30-45 minutes**.

3. **Sample Preparation**: Mix the protein samples with an appropriate volume of **2X SDS-PAGE sample buffer**. If working with reducing conditions (for denaturing disulfide bonds), add **β-mercaptoethanol** to a final concentration 5% v/v. **Heat the samples in a boiling water bath for 5 minutes** to denature the proteins.

4. **Loading and Electrophoresis**: Carefully **remove the comb from the gel** and clean any remaining acrylamide gel debris. Rinse the wells with the running buffer. **Load the protein samples into the wells, along with a molecular weight marker for reference**. Connect the electrophoresis apparatus to the power pack. **Run the gel at a constant voltage** (e.g., 120V) until the tracking dye reaches the bottom of the gel. This typically takes **1-2 hours**.

5. **Gel Staining and Destaining**: After electrophoresis, **carefully remove the gel** from the apparatus and **place it in a plastic container**. Stain the gel with **Coomassie Brilliant Blue R250 solution** (e.g., 0.1% w/v in 40% methanol, 10% acetic acid) for **at least 1 hour with gentle agitation. Destain the gel by washing it with a destaining solution** (e.g., 40% methanol, 10% acetic acid) until the protein bands are visible against a colorless background.

6. **Documentation and Analysis**: Document the stained gel using an **imaging system or a gel documentation system**. Analyze the protein bands using appropriate software, comparing their migration distances with the molecular weight marker to determine their sizes.

Native PAGE

Native Polyacrylamide Gel Electrophoresis (PAGE) is a powerful technique used to separate proteins **based on their native charge, size**, and **shape**. It allows researchers to analyze the intact, non-denatured protein complexes in their native conformation. This protocol focuses on **Blue Native PAGE**, which utilizes the anionic dye Coomassie Brilliant Blue to provide a negative charge on proteins, reducing their tendency to aggregate and **offering higher resolution for protein complexes ranging from 10 kDa to 10 MDa**.

Materials

1. Latex gloves
2. Weighing Balance

3. Graduated cylinder (200 ml)
4. Electrophoresis apparatus
5. Power pack
6. Pipettes
7. Eppendorf tubes
8. Boiling water bath
9. Magnetic stirring bar and stirrer
10. Beakers

Reagents

1. N, N, N', N'-Tetramethylethylenediamine (TEMED)
2. Ammonium Persulfate (APS)
3. Tris-HCl buffer
4. Glycine
5. Bromophenol Blue tracking dye
6. Coomassie Brilliant Blue R250
7. Acrylamide
8. Bis-Acrylamide
9. Glycerol

Protocol

1. **Prepare the Native PAGE Gel**: Commonly used percentages are **6-20%
depending on the protein size range**. Weigh the **appropriate amounts
Acrylamide and Bis-Acrylamide** using a weighing balance and transfer them
to a beaker. **Add running buffer (Tris-HCl)** to the beaker to reach the desired
gel volume. **Stir the mixture gently** using a magnetic stirrer until the reagents
are completely dissolved. **Add a small amount of Bromophenol Blue
tracking dye** to the gel mixture for visual tracking during electrophoresis. Pour
the gel mixture carefully into the electrophoresis apparatus, leaving about 1 cm
of space for the stacking gel.

2. **Prepare the Stacking Gel**: Prepare the stacking gel by **mixing the
appropriate amounts of Acrylamide, Bis-Acrylamide, and running buffer
in a beaker. Add a small amount of TEMED and APS** as catalysts and
initiators, respectively, to initiate the polymerization process. **Stir the mixture
gently** and **pour it on top of the resolving gel** in the electrophoresis apparatus.
Immediately insert a comb into the stacking gel to create wells for sample
loading.

3. **Sample Preparation**: **Mix the protein samples** of interest with a non-
denaturing, non-reducing sample buffer containing **glycerol to increase the**

sample density. **Load the protein samples into the wells** created in the stacking gel using pipettes.

4. **Electrophoresis**: Connect the electrophoresis apparatus to a power pack and **run the gel at a constant voltage** suitable for your protein of interest (usually 100-150 V) **until the dye front reaches the bottom of the gel** (typically 1-2 hours).

5. **Staining**: Prepare a staining solution **by dissolving Coomassie Brilliant Blue R250 in a mixture of methanol and acetic acid**. Remove the gel from the electrophoresis apparatus and place it in the staining solution. Gently shake the staining solution for about 1-2 hours or until the protein bands are sufficiently stained.

6. **Destaining**: Transfer the stained gel to **a destaining solution containing methanol and acetic acid without Coomassie Brilliant Blue R250**. Gently shake the destaining solution until the background is clear, and protein bands become more visible.

7. **Documentation and Analysis**: Carefully **remove the gel from the destaining solution** and **rinse it with water**. Place the gel on a transparent surface and **use a gel documentation system or a scanner to capture the protein band patterns**. Analyze the gel images to study the protein complexes' native structures, sizes, and interactions.

Agarose gel electrophoresis

Agarose gel electrophoresis is a widely used technique **to separate DNA molecules based on their size**. This technique is essential for various applications in molecular biology, such as **DNA fragment analysis, PCR product verification**, and **DNA purification**.
Fundamental concept:
   Agarose gel electrophoresis is a fundamental technique in molecular biology used to separate DNA fragments based on their size. It involves **three main steps**:
1. **Gel Preparation**: Agarose, a polymer extracted from seaweeds, is dissolved by boiling and then poured into a mold to solidify. As it cools, the agarose undergoes polymerization, forming a gel with pores whose size is determined by agarose concentration.

2. **Electrophoresis of DNA Fragments**: DNA is negatively charged at neutral pH, and when an electric field is applied across the gel, DNA molecules move

toward the positively charged anode. The migration of DNA through the gel depends on its size, agarose concentration, DNA conformation, and the applied current. **Smaller DNA fragments move faster through higher agarose concentrations, while larger fragments move more easily through lower agarose concentrations.**

3. **Visualization of DNA Fragments**: Since DNA is not naturally colored, it is stained with a dye specific to DNA after electrophoresis. The stained DNA appears as discrete bands on the gel. Alternatively, a fluorescent dye like **ethidium bromide** can be added to the agarose gel, and the DNA bands are **visualized under UV light** as they fluoresce.

To monitor the progress of gel electrophoresis, **tracking dyes** with known migration speeds are loaded along with each DNA sample. These dyes move through the gel and indicate the completion of the run when they reach the anode.

Virtual learning aid

Materials and reagents

1. Horizontal gel electrophoresis apparatus
2. UV transilluminator or gel documentation system
3. 50X TAE buffer (pH 8.5)
4. Agarose powder
5. Tris base
6. Glacial acetic acid
7. EDTA
8. Ethidium bromide solution (EtBr)
9. Distilled water
10. Gel loading dye (Bromophenol blue, Xylene cyanol, Sucrose)
11. DNA samples
12. Molecular weight marker

Protocol

1. **Preparing Agarose Gel**: In a heat-resistant container, **add 0.8 grams agarose powder to 100 ml of 1X TAE buffer (0.8% Agarose gel)**. Place the container in a microwave or hot plate and heat until the agarose completely dissolves to form a clear solution. While the agarose is still warm, carefully **pour it into the gel casting tray. Add 10 ml of ethidium bromide solution to the gel** and gently swirl to mix evenly. **Allow the gel to solidify by leaving it undisturbed for about 20-30 minutes**. Once solidified, gently remove the comb from the gel to create wells for sample loading.

2. **Assembling Gel Electrophoresis Apparatus**: Place the gel into the gel tank, **ensuring that the wells are closer to the negative electrode (cathode). Pour enough 1X TAE buffer** into the gel tank until the gel is fully submerged.

3. **Preparing Samples and Loading: Mix 5 ml of each DNA sample with 1 ml of gel-loading dye** in separate microcentrifuge tubes. The loading dye helps track the migration of DNA during electrophoresis. Load the samples into the wells, using a micropipette. **Be gentle to prevent damaging the gel.**

4. **Adding Molecular Weight Marker**: Load **at least 1 or 2 wells** with uncut, **good quality DNA (50 ng and 100 ng) as molecular weight standards.** These molecular weight markers help determine the size of the DNA fragments in the test samples.

5. **Electrophoresis**: Connect the gel tank to a power supply and **set the voltage to 70 V**. Run the gel until the dye in the loading dye has migrated approximately one-third of the distance through the gel. **This should take around 30-40 minutes.**

6. **Staining the Gel**: Prepare a staining solution of ethidium bromide by **diluting 10 mg of EtBr in 1 ml of sterile distilled water. Submerge the gel in the staining solution for approximately 15-30 minutes**, ensuring complete coverage.

7. **Destaining the Gel**: Transfer the gel to a clean container and add enough distilled water to fully submerge the gel. **Allow the gel to destain for about 15-30 minutes**, or until the background is clear, and DNA bands are visible.

8. **Visualizing and Documenting the Gel**: Place the gel back on the **UV transilluminator** or in the **gel documentation system**. Turn on the UV light to **visualize the EtBr-stained DNA bands**. Carefully document the gel image using a camera or the gel documentation system.

Safety and Disposal

Safely dispose of the agarose gel and staining solution according to your institution's guidelines and local regulations for hazardous waste disposal.

Note

Isoelectric focusing: Isoelectric focusing (IEF) is an electrophoretic technique used to separate proteins based on their isoelectric points (pIs), which is the pH at which a protein molecule carries no net charge. The principle behind IEF is that proteins are positively charged in solutions with pH values lower than their pI (pH < pI) and negatively charged in solutions with pH values higher than their pI (pH > pI). When an electric field is applied to a pH gradient, proteins move towards the position in the gradient where their net charge becomes zero, resulting in the formation of distinct bands at their characteristic pI values.

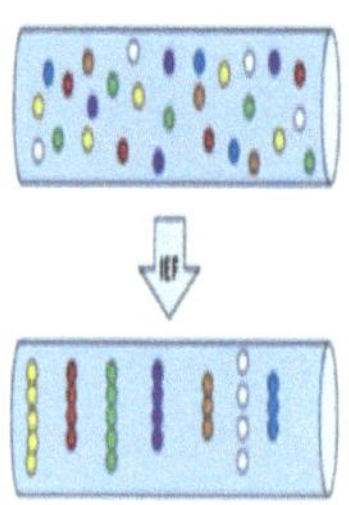

Two-dimensional gel electrophoresis

Two-dimensional (2D) gel electrophoresis is a widely used technique to separate complex protein mixtures **based on their isoelectric point (pI)** and **molecular weight**. It involves **two steps: isoelectric focusing (IEF)** and **sodium dodecyl sulfate-polyacrylamide gel electrophoresis (SDS-PAGE)**. IEF separates proteins based on their charge, while SDS-PAGE further separates proteins based on their size. The resulting 2D gel provides a protein profile, which can be used for various applications such as **protein identification**, comparison of **protein expression levels**, and **post-translational modification analysis**.

Fundamental concept:

Two-dimensional gel electrophoresis is based on separating a mixture proteins according to **two properties**, one in each dimension. The first dimension separates proteins according to their **native isoelectric point (pI)** using Isoelectric focusing and the second dimension separates by mass using **SDS PAGE**.

First dimension separation

This is performed by **denaturing IEF**. Using this proteins are separated **based on their pI**. In IEF, proteins are electrophoresed into a pH gradient. As the proteins move through the gradient, they encounter **a point where the pH is equal to their pI and they stop migrating**. Because of the difference in pI, different protein stops at a different point in the gradient. A conditioning step is applied to proteins separated by IEF before the second dimension separation. This process reduces disulfide bonds and alkylates the resultant sulfhydryl groups of cysteine residues. Concurrently, proteins are coated with SDS for separation based on mass.

Second dimension separation

This part is performed by **SDS-PAGE**. Proteins that have been separated on an IEF gel can then be separated in a second dimension **based on their size or mass**. To accomplish this, the IEF gel is placed lengthwise on a second PAGE saturated with SDS. When the electric field is imposed, the proteins migrate from the IEF gel into the SDS gel and then separate according to their mass. The sequential resolution of proteins by their charge and mass can achieve excellent separation of cellular proteins.

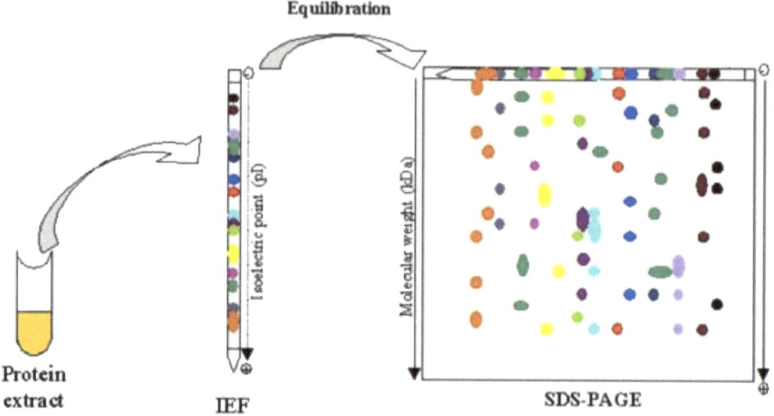

Reagents

1. Immobilized pH gradient (IPG) strips (pH range determined by your experiment)
2. IEF Sample buffer (urea, thiourea, CHAPS, dithiothreitol (DTT), carrier ampholytes, and a trace amount of bromophenol blue)
3. Mineral oil (for covering IPG strips during rehydration)
4. Equilibration buffer (Tris-HCl, urea, glycerol, SDS, and bromophenol blue)
5. 12% polyacrylamide gel for SDS-PAGE
6. 10% SDS running buffer
7. Coomassie or silver stain solution
8. Destaining solution (if using Coomassie stain)
9. Protein molecular weight marker

Equipment

1. Isoelectric focusing (IEF) apparatus
2. SDS-PAGE gel apparatus
3. Power supply for electrophoresis
4. pH meter and electrode
5. Protein sample preparation equipment (homogenizer, centrifuge, etc.)
6. Gel imaging system (Coomassie-stained gels) or silver staining setup
7. Protein quantification kit (e.g., Bradford assay)

Protocol

1. **Protein Sample Preparation: Harvest and lyse the cells or tissues** interest in a suitable **lysis buffer containing protease inhibitors. Sonicate or**

homogenize the lysate to ensure complete cell/tissue disruption and protein extraction. **Centrifuge the lysate** at a suitable speed and time to remove cell debris and **collect the supernatant** (total protein extract). Determine the protein concentration using a protein quantification kit like the **Bradford assay**.

2. **Rehydration of IPG Strips**: Rehydrate the IPG strips **in the IEF Sample buffer containing the total protein extract**. Add the appropriate volume sample to the buffer (as per the manufacturer's guidelines) to achieve the desired protein load.

3. **Isoelectric Focusing (IEF)**: Place the rehydrated IPG strips on the **IEF apparatus**, ensuring **proper contact with the electrode strips**. Cover the strips with a **thin layer of mineral oil** to prevent evaporation during the run. Run the IEF under conditions appropriate for your experiment, including the desired pH gradient range and voltage settings.

4. **Equilibration of IPG Strips**: After the IEF run, carefully **remove the mineral oil from the strips**. Equilibrate the IPG strips in **two steps**: **Incubate the strips in equilibration buffer I** (without DTT) for about 15 minutes. **Transfer the strips to equilibration buffer II** (containing DTT) for another 15 minutes.

5. **SDS-PAGE**: Place the equilibrated IPG strips **on top of the SDS-PAGE gel**, ensuring proper contact. Fill the SDS-PAGE gel apparatus with a **10% SDS running buffer**, covering the gel and the IPG strips. Run the SDS-PAGE under appropriate conditions, typically at a constant voltage.

6. **Staining and Visualization**: After SDS-PAGE, remove the gel from the apparatus and stain the gel using **Coomassie or silver stain** following standard protocols. If using Coomassie staining, destain the gel to reduce background staining. Visualize and document the stained gel using a gel imaging system or other suitable equipment.

Pulse field gel electrophoresis

In **conventional gel electrophoresis**, the size of DNA fragments that can be separated is **limited to around 50 kilobases (kb)**. This constraint arises because very large DNA fragments are unable to penetrate the pores in the agarose gel, making it challenging to resolve them effectively. To overcome this limitation, a groundbreaking technique called **Pulse Field Gel Electrophoresis (PFGE)** was invented. PFGE allows the **separation of DNA fragments of up to 10 megabases (Mb)** in size. This technique involves electrophoresis in an agarose gel, where **two electric fields are applied alternately at different angles for specific periods**. By applying this innovative method, scientists can elongate DNA molecules in the direction of the electric field and achieve enhanced separation of large DNA fragments.

Fundamental concept

When an electric field is applied to the agarose gel during PFGE, the DNA molecules undergo elongation in the direction of the electric field. Following this initial field, a second electric field is applied. For the DNA to migrate in the direction of this new field, it must change its conformation and reorient itself. Provided that the alternating fields have equal voltage and pulse duration, the DNA will migrate in a straight path down the gel. This unique method allows researchers to effectively separate large DNA fragments that were previously challenging to resolve.

Modern PFGE instruments utilize **multiple electrodes arranged in a hexagonal array**. A sophisticated variant of PFGE, known as **Contour-Clamped Homogenous Electric Fields (CHFE)**, employs some of the **electrodes that are clamped or held at intermediate potentials**. This clamping action results in homogenous electric fields, which are essential for producing distortion-free lanes during the electrophoretic separation.

Reagents

1. Agarose
2. TBE Buffer
3. Ethidium Bromide

Equipment

1. PFGE Electrophoresis Apparatus
2. Gel Casting Tray and Comb
3. Power Supply
4. UV Transilluminator

Protocol

1. **Prepare the Agarose Gel**: Measure the appropriate amount of agarose and TBE buffer according to the desired gel percentage. Heat the mixture in a microwave or on a hot plate until the agarose is completely dissolved. Allow the mixture to cool slightly before pouring it into the gel casting tray. Insert the comb into the gel, creating wells for loading DNA samples. Let the gel solidify.

2. **DNA Sample Preparation**: Isolate the DNA of interest using standard extraction methods. **Treat the DNA with appropriate restriction enzymes to digest it into fragments**.

3. **Loading and Electrophoresis**: Carefully load the digested DNA samples into the wells of the agarose gel. Fill the electrophoresis chamber with TBE buffer. Place the gel into the chamber, ensuring that it is fully submerged in the buffer. Connect the gel to the power supply and **apply the alternating electric fields at defined angles and time periods**.

4. **Visualization of DNA Bands**: After electrophoresis, carefully remove the gel from the chamber. Stain the DNA with **ethidium bromide for approximately 30 minutes**. Visualize the DNA bands under a **UV transilluminator**.

Capillary electrophoresis

Capillary electrophoresis (CE) is a powerful analytical technique that utilizes **narrow-bore capillaries** (typically 20-100 μm in diameter) to achieve high-efficiency separation of molecules, both large and small. This process is facilitated **by applying high voltage**, which generates **two distinct flows within the capillary: electroosmotic flow** of buffer solutions and the **electrophoretic flow** of ionic species. The main driving force behind CE is electroosmosis, which occurs due to the **ionization of silanol groups** on the inner walls of fused silica capillaries when exposed to a buffer with a pH greater than approximately 2 or 3. The resulting **negatively charged silanate ions** (SiO^-) create fixed and mobile cation layers on the capillary walls. When a voltage is applied, the mobile cation layer moves towards the cathode, causing the bulk buffer solution to migrate along with it, resulting in electroosmotic flow.

Virtual learning aid

Reagents and Equipment

1. **Running Buffer**: A buffer solution with a specific pH to control the extent of electroosmosis. The pH affects the ionization state of silanol groups on the capillary walls, which in turn influences electroosmotic flow.

2. **Sieving Matrix**: Fused silica capillaries are filled with a sieving matrix, which can be a chemically cross-linked gel (e.g., polyacrylamide) or a flowable polymer (e.g., modified cellulose or non-cross-linked polyacrylamide). This matrix helps separate DNA fragments based on their size, similar to standard slab gel electrophoresis.

3. **High Voltage Power Supply**: To generate the necessary high voltage for the electrophoretic flow of ionic species within the capillary.

4. **Capillary Electrophoresis Instrument**: The main device used for CE, which consists of the narrow-bore capillary, a detector, and a data handling system.

Protocol

1. **Preparation of Running Buffer**: Prepare the running buffer with the desired pH (based on the analysis requirements) using **appropriate buffer salts** and adjust the pH using acid or base.

2. **Preparation of Sieving Matrix**: Fill the fused silica capillary with the selected **sieving matrix**. For cross-linked gels, use **chemical cross-linking agents** following the manufacturer's instructions.

3. **Sample Preparation**: Prepare your DNA samples by mixing them with a suitable **loading buffer**, which contains **tracking dyes** to monitor the migration of the DNA fragments.

4. **Capillary Conditioning**: Flush the capillary with the **running buffer** to remove any impurities or bubbles and ensure proper equilibration.

5. **Sample Loading**: Inject the prepared DNA samples into the capillary using appropriate injection methods, such as **pressure or electrokinetic injection**.

6. **Applying Voltage**: Apply the desired voltage across the capillary using the high-voltage power supply. The positively charged anode and negatively charged cathode will create an electrophoretic flow of DNA fragments toward the anode.

7. **Detection and Data Analysis**: Monitor the migration of DNA fragments using the detector integrated into the capillary electrophoresis instrument. The data handling system will process the results and provide you with valuable insights into the size and quantity of DNA fragments.

Note

1. In electrophoresis, there are two common types of gels used for separating molecules: Polyacrylamide gel and agarose gel. These gels act like molecular sieves, and their ability to separate molecules depends on the size of the gel pores. Agarose gel is ideal for separating large nucleic acids due to its large pore size, enabling the separation of DNA ranging from about 50 to 50,000 base pairs (bp). On the other hand, polyacrylamide gel has a smaller pore size, making it perfect for separating smaller nucleic acid molecules within the range of about 5 to 3000 bp. The resolving power of agarose gel is between 5 to 10 nucleotides, while polyacrylamide gel can achieve a single-nucleotide separation range.

2. Electrophoresis buffers play a crucial role in separating DNA and RNA during agarose and polyacrylamide gel electrophoresis. The right buffer choice depends on factors like sample size, run time, and post-electrophoresis processes. For nucleic acids, the commonly used running buffers are TAE (Tris-acetate-EDTA) and TBE (Tris-Borate-EDTA). TBE has higher ionic strength and is ideal for separating small samples (<1 kb), while TAE is recommended for larger fragments (>12 kb) due to its lower buffering capacity.

During gel preparation, a gel casting buffer with suitable electrical conductivity enables nucleic acid mobility. Before loading the gel, samples are mixed with a gel loading buffer containing density gradient agents (glycerol or sucrose), salts (Tris-HCl), and metal chelators (EDTA) to prevent nucleic acid degradation. Loading dyes are also added to help monitor the process visually. The ideal dye should not interact with DNA or the gel, possess a negative charge to move towards the anode, provide color to the DNA sample, and have a size within the gel's resolution limits. Bromophenol blue is a common choice, serving as a tracking dye, and its migration signals the end of electrophoresis.

For separating single-stranded nucleic acids like RNA with complex secondary structures, denaturing buffers containing agents like urea or formamide are used. Denaturing electrophoresis is especially routine for RNA analysis. Typical denaturing buffers include glyoxal and DMSO in sodium phosphate buffer or formamide in MOPS buffer for agarose gels, and urea in TBE buffer for polyacrylamide gels to maintain single-strandedness of the nucleic acids.

After completing the electrophoresis run, we need to analyze the protein bands either quantitatively or qualitatively. There are two main ways to visualize the separated protein bands in the gel. One approach is using Western blotting, where the proteins are transferred to a membrane for further analysis. Alternatively, we can directly visualize the proteins within the gel using various staining or detection methods.

One commonly used dye for protein visualization is **Coomassie brilliant blue**, which selectively binds to proteins rather than the gel itself. It has a low detectable limit, making it capable of detecting about 0.1-0.5 µg of protein. The dye binds non-specifically to almost all proteins, making it widely applicable for visualization purposes.

On the other hand, **silver staining** is the most sensitive method for permanent and visible protein staining in polyacrylamide gels. In this technique, silver ions are reduced to metallic silver on the protein bands, resulting in a distinct brown-black color. Silver ions specifically interact and bind to certain amino acid groups, such as carboxylic acid groups (Asp and Glu), Imidazole (His), sulfhydryls (Cys), and amines (Lys). This method is remarkably sensitive, as it can detect proteins at an incredibly low limit of about 0.5 ng, making it a preferred choice for highly sensitive protein visualization.

Virtual learning aid

Estimation of molecular weight of protein

SDS-PAGE is a method for estimating the molecular weight of unknown proteins. It involves comparing the migration distance of the unknown protein through a gel with that of proteins of known molecular weight.

The electrophoretic mobility of a protein coated with SDS on the gel is inversely proportional to the logarithm of its molecular weight. To determine the molecular weight of an unknown protein, a set of standard proteins with known molecular weights (molecular weight markers) is run alongside the samples in the same gel. These molecular weight markers are separated into bands, serving as a reference for determining the molecular weight of the unknown sample proteins.

The determination of molecular weight involves separating the sample and molecular weight markers on the same gel. The relative migration distance (R) is calculated as the migration distance of a protein band divided by the migration distance of the dye front, which is normalized to the tracking dye that migrates slightly behind the ion front. Using the obtained values for the molecular weight markers, a graph is plotted to correlate the log of the molecular weight of SDS-denatured polypeptides and their relative migration distance (R). A linear plot indicates the full denaturation of proteins and an appropriate gel percentage for the molecular weight range. However, at extreme molecular weight values, the standard curve becomes sigmoid due to the sieving effect of the gel matrix, hindering penetration at high molecular weight and allowing almost free migration at low molecular weight. By interpolating from this graph, the molecular weight of the unknown protein band can be determined.

Virtual learning aid

Chapter 5

Chromatography

Introduction

Chromatography is a powerful method used **to separate different compounds from a mixture**. Its origins can be attributed to the work of the **Russian botanist Mikhail Tswett**, who first applied the technique **to separate plant pigments** like chlorophyll and anthocyanins. He accomplished this by passing solutions of these compounds through a glass column filled with finely divided calcium carbonate. As the compounds traveled through the column, they appeared as distinct colored bands, leading to the method being named chromatography (from the **Greek words chroma**, meaning "color," and **graphein**, meaning "writing").

The underlying principle of chromatography is straightforward: the sample, referred to as the **solute or analyte, interacts with both a stationary phase and a mobile phase**. The two phases can be of different physical states, such as a solid and a liquid, a gas and a liquid, or a solid and another fluid. The **stationary phase remains fixed**, while the **mobile phase carries the sample** through it. Depending on the specific chromatographic method employed, the mobile phase may be a liquid (liquid chromatography) or a gas (gas chromatography). Regardless of the variation, all chromatographic techniques involve the passage of a mobile phase through a stationary solid phase. This arrangement allows different components of the sample to distribute themselves between the mobile and stationary phases to varying degrees, thus leading to separation.

Classification of chromatographic methods

1. Based on the **shape** of the chromatographic bed:
    A. **Planar chromatography**
    B. **Column chromatography**

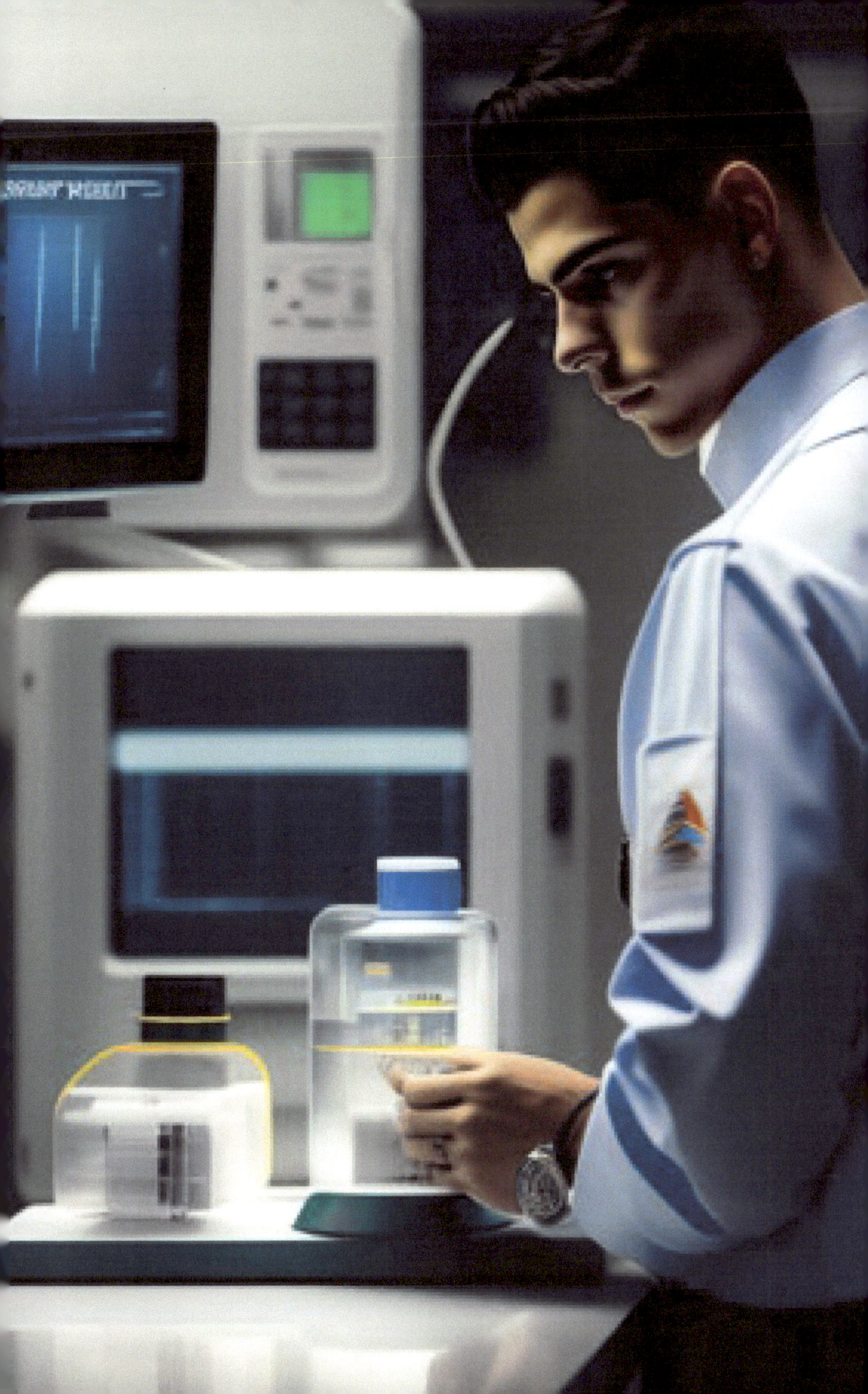

2. Based on the **physical state** of the mobile and stationary phases:
- A. Liquid chromatography
- B. Gas chromatography
- C. Supercritical fluid chromatography

3. Based on the **mechanism** of separation:
- A. Adsorption chromatography
- B. Partition chromatography
- C. Ion exchange chromatography
- D. Size exclusion chromatography

Virtual learning aid

Based on the shape of the chromatographic bed

A. Planar chromatography is a technique where the stationary phase is applied on a flat surface. This surface can either be a paper treated with a substance serving as the stationary phase, known as **paper chromatography (PC)**, or a thin layer of a substance acting as the stationary phase spread on a glass, metal, or plastic plate, known as **thin-layer chromatography (TLC)**. Paper chromatography is also referred to as open-bed chromatography. In thin-layer chromatography, silica gel or alumina is commonly used as the stationary phase, with silica gel being the most frequently employed. Silica gel has a structure where silicon atoms are bonded together through oxygen atoms in a

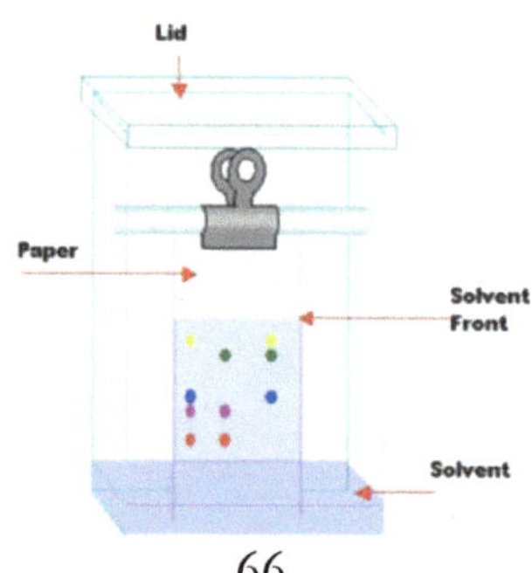

large covalent network. The alternative stationary phase, alumina (Aluminium oxide), is also used in certain cases.

Paper chromatography

Paper chromatography is employed for separating pigments in plants, quality control in pharmaceuticals, food analysis, forensic ink and dye analysis, environmental testing, clinical analysis of body fluids, separation of amino acids and proteins, chemical education, detecting counterfeit products, drug testing, analyzing plant extracts, and textile dye analysis.

Virtual learning aid

Materials

1. Paper chromatography sheets or filter paper
2. Sample to be analyzed
3. Solvent (mobile phase)
4. Developing chamber (glass jar with lid)
5. Capillary tubes or micropipettes
6. Ruler
7. Pencil
8. UV lamp or suitable visualization method (e.g., iodine chamber)

Protocol

1. **Prepare the sample**: Dissolve or suspend the sample in a small volume of a suitable solvent to create a concentrated solution.

2. **Prepare the chromatography sheet**: Draw a horizontal baseline near one end of the paper using a pencil.

3. **Apply the sample**: Using a capillary tube or micropipette, spot a small volume of the sample solution onto the baseline. Allow it to dry.

4. **Prepare the developing chamber**: Pour a small amount of the chosen solvent into the developing chamber, enough to cover the bottom by about 1-2 cm.

5. **Start the chromatography**: Gently place the paper in the chamber, ensuring the baseline is above the solvent level, and close the lid to allow the chromatogram to develop.

6. **Visualization**: Once the solvent front reaches the desired height (usually about 2/3 of the paper), remove the paper from the chamber and mark the solvent front. Let the paper dry.

7. **Visualize the chromatogram**: Examine the dried paper under a UV lamp or use an iodine chamber to detect separated compounds as spots along the paper strip.

Thin layer chromatography (TLC)

Thin Layer Chromatography (TLC) is employed for compound identification, purity assessment, monitoring reactions, and separation of mixtures. It finds

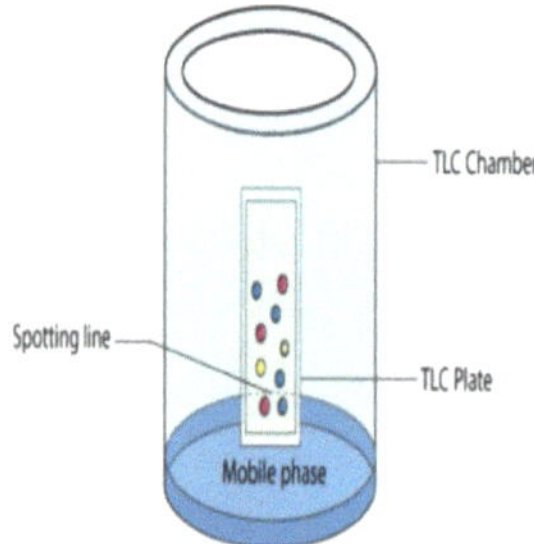

applications in fields such as pharmaceuticals, natural products, forensics, environmental analysis, and food testing.

Virtual learning aid

Scan It!

Materials

1. TLC plates (silica gel or other suitable material)
2. Sample to be analyzed
3. Solvent system (mobile phase)
4. Developing chamber (with a lid)
5. Capillary tubes or micropipettes
6. UV lamp or other visualization method

Protocol

1. **Prepare the sample**: Dissolve the sample in a suitable solvent to create a concentrated solution.

2. **Prepare the TLC plate**: Draw a horizontal baseline near one end of the plate using a pencil.

3. **Apply the sample**: Using a capillary tube or micropipette, spot a small volume of the sample solution onto the baseline. Allow it to dry.

4. **Prepare the developing chamber**: Pour a small amount of the solvent system into the developing chamber to a depth of about 1-2 cm.

5. **Start the chromatography**: Place the TLC plate in the chamber, ensuring the baseline is above the solvent level, and close the lid.

6. **Visualization**: Allow the solvent to move up the plate, carrying the sample components. Once the solvent front reaches the desired height, remove the plate from the chamber and mark the solvent front. Let the plate dry.
7. **Visualize the chromatogram**: Examine the dried TLC plate under UV light or other suitable visualization methods to detect separated spots corresponding to different components of the sample.

Column chromatography

Column chromatography is a separation technique used in chemistry, **where a tube is filled with a stationary phase**. This stationary phase can be in the form of solid particles coated with a liquid, and it may either fill the entire tube (packed column) or be concentrated on the inner tube wall, leaving an open pathway for the mobile phase in the middle (open-tubular column).

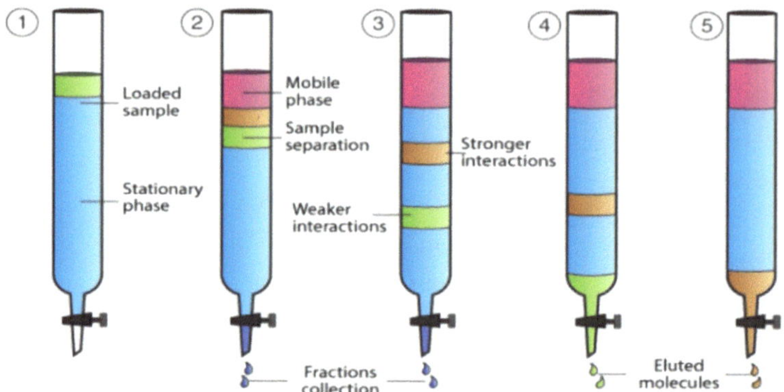

Virtual learning aid

Materials

1. Column (glass or plastic) packed with suitable stationary phase material (e.g., silica gel, Sephadex, or alumina)
2. Sample to be separated
3. Solvent system (mobile phase)
4. Glass wool or cotton to form a solvent reservoir
5. Stopcock or clamp to control the flow of solvent

Protocol

1. **Column preparation**: Choose a column size appropriate for your sample volume and column material suitable for your separation. **Pack the column with the stationary phase**, ensuring there are **no air bubbles and a uniform packing**.

2. **Sample loading**: Prepare the sample by dissolving it in a small volume of a suitable solvent. Load the sample **onto the top of the column**.

3. **Elution**: Add the solvent system to the column slowly to allow the solvent to flow through the column. **Collect the eluted fractions in separate containers**.

4. **Fraction collection**: Collect fractions based on the elution times of different components. Monitor the elution using appropriate detection methods (e.g., UV detector).

5. **Analysis**: Analyze the collected fractions using suitable analytical techniques (e.g., TLC or HPLC) to identify and quantify the separated components.

Based on the physical state of the mobile and stationary phases

Chromatography, a powerful analytical technique, can be classified into two main types: **gas chromatography** and **liquid chromatography**. In gas chromatography, the separation is based on the physical properties of the mobile phase, and it can be further divided into two subgroups: **gas-solid chromatography** and **gas-liquid chromatography**, depending on the nature of the stationary phase. Similarly, in liquid chromatography, the separation is also determined by the physical properties of the mobile phase, and it can be categorized into two subgroups: **liquid-solid chromatography** and **liquid-liquid chromatography**, based on the nature of the stationary phase. These distinct chromatographic techniques play essential roles in various scientific applications.

High-performance liquid chromatography

High-Performance Liquid Chromatography (HPLC) is an advanced version of **liquid column chromatography** that offers significant improvements. Instead of relying on gravity-driven solvent flow through a column, HPLC **employs high pressures of up to 400 atmospheres to force the solvent through the column**. This enhancement results in much faster separation. Furthermore, HPLC allows the use of small-sized particles as column packing material, leading to a substantially larger surface area for interactions between the stationary phase and the flowing molecules. This increased surface area enables superior separation of mixture components. The key advantages on HPLC stem from the development of stationary supports with **small particle sizes and large surface areas**, as well as the application of high pressure to the solvent flow, which improves elution rates.

HPLC is a liquid chromatographic technique where the mobile phase consists of a liquid rather than a solvent allowed to pass through a column under gravity. By pumping the mobile phase under high pressure, HPLC achieves exceptional performance and speed compared to traditional column chromatography. **HPLC operates in various chromatographic modes, including adsorption, partition, ion exchange, and size exclusion, often leading to combined separation effects**. When the stationary phase is solid, it is referred to as adsorption chromatography, while with a liquid stationary phase, the sample undergoes partition chromatography. **Reversed-phase HPLC**, the most commonly used form, utilizes a non-polar stationary phase and a polar mobile phase. The **efficiency of an HPLC column** is determined **by its packing material's nature**, **type**, and **size**, as well as the column dimensions. In this context, resolution, which signifies how well solutes are separated, serves as a crucial parameter. HPLC's enhanced resolution, compared to classical column chromatography, is primarily attributed to the use of adsorbents with small particle sizes and large surface areas. The application of high pressure combined with small particle adsorbents enables HPLC to achieve **high resolving power and short analysis time**, making it a preferred chromatographic technique.

Virtual learning aid

Advantages of HPLC

1. **High Speed**: Analysis times are impressively quick, often measured in minutes or even seconds, allowing for rapid sample processing and data generation.

2. **High Resolution**: HPLC provides exceptional separation power due to columns densely packed with small, uniform particles, resulting in clear and distinct separation of complex mixtures.

3. **High Sensitivity**: HPLC can detect trace amounts of substances with remarkable precision, achieving detection limits ranging from parts-per-million (ppm) to sub-parts-per-billion (ppb), making it ideal for analyzing low-concentration samples.

4. **High Accuracy**: The combination of high-precision sampling devices and reliable standards ensures accurate and reliable analysis results, instilling confidence in the obtained data.

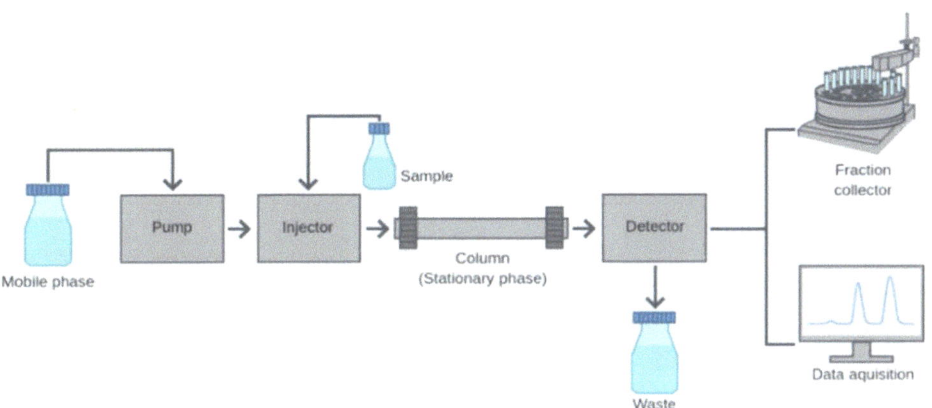

Materials

1. HPLC system (pump, injector, column, detector)
2. Sample to be analyzed
3. Mobile phase and buffers
4. Solvent reservoirs
5. Fraction collector (optional)

Protocol

1. **Column selection**: Choose an appropriate HPLC column based on the analytes and separation requirements.

 2. **Sample preparation**: Prepare the sample by dissolving it in a suitable solvent and filtering it to remove any particulates.

3. **Mobile phase selection**: Prepare the mobile phase and optimize the solvent composition and gradient program if required.

4. **HPLC setup**: Assemble the HPLC system and equilibrate the column with the mobile phase.

5. **Sample injection**: Inject the prepared sample into the HPLC system using an autosampler or manual injection.

6. **Elution**: Start the HPLC run and allow the mobile phase to carry the sample through the column.

7. **Detection**: Monitor the elution using a suitable detector (e.g., UV-Vis, fluorescence, or MS) to detect and quantify the separated compounds.

8. **Data analysis**: Analyze the chromatograms using appropriate software to identify and quantify the analytes of interest.

Gas chromatography

Gas chromatography (GC) is a widely used analytical technique **for analyzing volatile substances in the gas phase**. It involves **two phases**: a **mobile phase consisting of a carrier gas** and a **stationary phase**. The carrier gas, which must be dry, oxygen-free, and chemically inert, serves to transport the analytes through the system. There are different types of carrier gases such as **helium, nitrogen, argon**, and **hydrogen**, chosen based on performance requirements and the type of detector used. The stationary phase can either be a solid

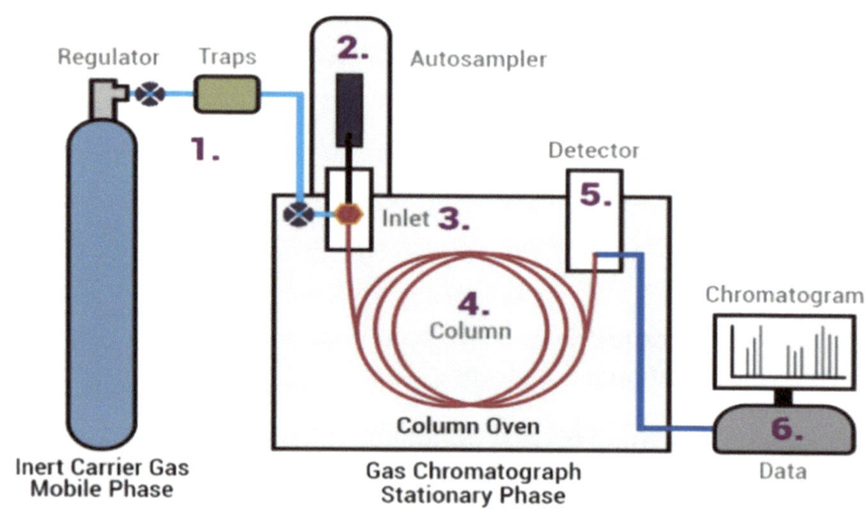

adsorbent (**gas-solid chromatography, GSC**) or a liquid on an inert support (**gas-liquid chromatography, GLC**). In GSC, retention of analytes occurs due to physical adsorption on the solid surface, while in GLC, the analyte is partitioned between the gaseous mobile phase and the liquid phase immobilized on an inert solid packing or capillary tube. GC is ideal for analyzing thermally stable, volatile substances and stands out from other chromatography methods because it does not rely on the mobile phase for interacting with the analyte. To control the carrier gas flow rate, pressure regulators, gauges, and flow meters are used along with pressurized tanks containing the selected carrier gas.

Virtual learning aid

Note

Gas chromatography (GC) and liquid chromatography (LC) are analytical techniques used to separate and analyze mixtures of compounds. They differ in key parameters governing their separation processes. In GC, the nature of the stationary phase and temperature are the primary factors affecting separation. On the other hand, LC relies on the nature of both the stationary phase and the mobile phase, and to some extent, temperature, for achieving separation. Temperature plays a crucial role in both techniques, but it is even more critical in GC. Moreover, in LC, both the mobile phase and stationary phase interact with the analytes to facilitate separation, while in GC, only the stationary phase is involved in this process. Understanding these differences is essential for selecting the appropriate chromatographic method for a given analysis.

Materials

1. GC system (injector, column, oven, detector)

2. Sample to be analyzed
3. Carrier gas (e.g., helium or nitrogen)
4. Syringe or autosampler for injection
5. Gas chromatography standards (optional)
6. Data analysis software (optional)

Protocol

1. **Sample preparation**: Prepare the sample by dissolving it in a suitable solvent or extracting it from the matrix (if required).

2. **Injection**: Load the prepared sample into a syringe or use an autosampler to inject it into the GC system.

3. **Column selection**: Choose an appropriate GC column based on the analytes and separation requirements.

4. **Oven temperature program**: Set the temperature program for the GC oven to optimize the separation.

5. **Carrier gas flow**: Set the carrier gas flow rate to ensure proper sample transport through the column.

6. **Elution**: Start the GC run and allow the carrier gas to carry the sample through the column.

7. **Detection**: Monitor the elution using a suitable detector (e.g., FID, TCD, or MS) to detect and quantify the separated compounds.

8. **Data analysis**: Analyze the chromatograms using appropriate software to identify and quantify the analytes of interest.

Based on the mechanism of separation

The chromatographic techniques use several types of mechanisms to separate analytes. Based on the mechanism of separation, chromatographic techniques can be **partition, adsorption, size exclusion, affinity,** and **ion exchange chromatography**.

Chromatographic techniques are commonly categorized into two main types: **partition** and **adsorption chromatography**, depending on how solutes interact with the stationary phase. These interactions can be further classified into two types, although many separation processes involve a combination of both. Adsorption occurs when the sample components attracted to the surface of a solid stationary phase, while absorption (referred to as partition by IUPAC) happens when the sample components diffuse into the interior of the stationary phase.

Adsorption chromatography

In **adsorption chromatography**, the separation of components in a mixture is **based on their differing degrees of adsorption to the stationary phase** present in the chromatography column. The adsorption process involves **weak non-covalent interactions**, such as ionic interactions, van der Waals forces, hydrogen bonding, and hydrophobic interactions, between the components on the mixture and the stationary phase. **Polar compounds tend to adsorb strongly to polar stationary phases, while non-polar compounds preferentially adsorb to non-polar stationary phases**. This results in polar components eluting later when using a polar stationary phase, and non-polar components eluting first. Conversely, the order is reversed when using a non-polar stationary phase.

Virtual learning aid

Note

1. Adsorption chromatography Separation is based mainly on differences between the adsorption affinities of the sample components for the surface on an active solid stationary phase.

2. Partition chromatography Separation is based mainly on differences between the solubilities of the sample components in the stationary phase (gas-liquid chromatography), or on differences between the solubilities of the components in the mobile and stationary phases (liquid-liquid chromatography).

Liquid-liquid partition chromatography

Liquid-liquid partition chromatography, also known as **partition chromatography**, separates substances **based on their partitioning between a stationary liquid phase and a mobile liquid phase**. The solute's solubility in the mobile phase determines its movement through the system. Substances more soluble in the mobile phase move rapidly, while those favoring the stationary phase are delayed.

There are **two types of partition chromatography**: **normal phase** and **reversed phase**.

In **normal-phase partition chromatography**, the **stationary phase is more polar than the mobile phase**. This leads to the elution of the least polar analyte first and the most polar one last. In contrast, **reversed-phase partition chromatography** involves a **mobile phase significantly more polar than the stationary phase**. As a result, the most polar solutes elute first, and the least polar ones elute last. A specific type of partition chromatography is paper chromatography. In this method, a paper is dipped into a solvent mixture consisting of aqueous and organic components. The solvent moves into the paper through capillary action, with the aqueous component binding to the paper's cellulose, forming the stationary phase. The organic component becomes the mobile phase. The migration rates of the substances being separated depend on their relative solubilities in the polar stationary phase and the non-polar mobile phase. The separation process relies on the partition coefficient of each solute. The nonpolar molecules move faster than the polar ones. The separation's progress is often expressed using the dimensionless term **Rf (Relative front)**, which is the **ratio of the distance traveled by the substance to the distance traveled by the solvent front. Paper chromatograms** can be developed using **either ascending or descending solvent flow**, with descending chromatography having the advantages of faster separation aided by gravity and enabling solvent runoff for quantitative separations of substances with very small Rf values.

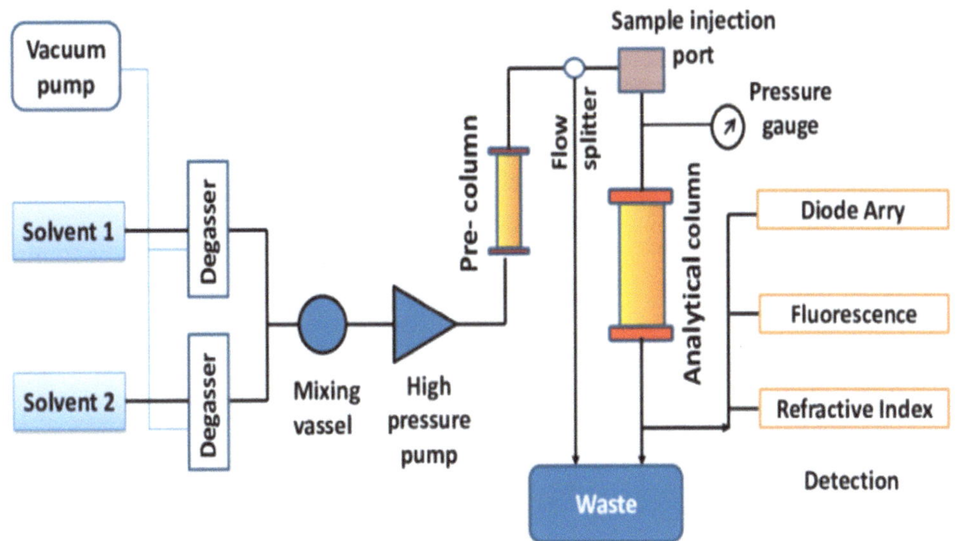

Virtual learning aid

Materials

1. Separatory funnel
2. Sample to be analyzed
3. Two immiscible solvents (usually water and an organic solvent)
4. Glassware for sample mixing and separation

Protocol

1. **Sample preparation**: Prepare the sample in a suitable solvent that is miscible with both the immiscible solvents to be used in the chromatography.

2. **Solvent selection**: Choose two immiscible solvents and mix them in the separatory funnel. The sample will be partitioned between these two solvents during the chromatography.

3. **Partitioning**: Add the sample solution to the separatory funnel and shake gently to allow the partitioning to occur.

4. **Separation of phases**: Allow the two phases to separate, forming distinct layers.

5. **Fraction collection**: Carefully drain the lower layer (usually the organic phase) into a separate container, containing the separated components.

6. **Analysis**: Analyze the collected fractions using suitable analytical techniques (e.g., TLC or HPLC) to identify and quantify the separated components.

Size exclusion chromatography

Size exclusion chromatography, also known as **molecular sieve chromatography**, is a separation technique that separates molecules **based on their size and shape**. It involves a **column filled with porous gel beads**, typically made from insoluble and hydrated polymers like **polyacrylamide (Sephacryl or BioGel P)**, **dextran (Sephadex)**, or **agarose (Sepharose)**, serving as the stationary phase. The process includes **two main types**: **gel permeation chromatography**, which uses an organic mobile solvent, and **gel filtration chromatography**, which employs an aqueous mobile solvent for separating and characterizing molecules.

Fundamental concept

The principle behind size exclusion chromatography is straightforward. **When a solution containing molecules of different sizes is passed through the column, smaller molecules can enter the pores in the gel beads, while larger ones cannot.** As a result, **larger molecules move faster and elute first**, while smaller molecules have longer retention times due to their ability to access the pores. By applying a mixture of proteins to the column and washing it with a suitable buffer, the proteins elute in order of decreasing molecular size. The **gel's exclusion limit represents the molecular mass of the smallest molecule unable to penetrate the gel pores**. For instance, a common gel like Sephadex G-50 has an exclusion limit of 30,000 Da, meaning that molecules larger than this value pass through the column bed without entering the gel pores.

The **distribution coefficient K** is a parameter that depends on the molecule's size, with **K=0 indicating complete exclusion** from the pores and **K=1 indicating entry into the porous beads** with access to the inner solvent. **Intermediate sizes will have K values ranging from 0 to 1.**

List of Materials commonly used for porous gel beads in size exclusion chromatography

Material	Trade Name	Fractionation Range (Molecular mass in Da)
Dextran	Sephadex G-10	0-700
	Sephadex G-25	1000-5000
	Sephadex G-50	1500-30,000
	Sephadex G-75	3000-70,000
	Sephadex G-100	4000-150,000
	Sephadex G-150	5000-300,000
	Sephadex G-200	5000-800,000
Polyacrylamide	Bio-gel P-2	100-1800
	Bio-gel P-6	1000-6000
	Bio-gel P-60	3000-60,000
	Bio-gel P-150	15,000-150,000
	Bio-gel P-300	60,000-400,000
Agarose	Sepharose 2B	$2 \times 10^6 - 25 \times 10^6$
	Sepharose 4B	$3 \times 10^5 - 3 \times 10^6$
	Sepharose 6B	$10^4 - 20 \times 10^6$

***The molecular mass listed are for globular proteins.**

Size measurements by size exclusion chromatography

Size exclusion chromatography is a technique used **to determine the size of the solute molecules**. To characterize the column, we divide the **total volume (VT)** into three parts: the **void volume (V0)** outside the packing material, the **internal volume (Vi)** within the porous beads, and the volume occupied by the **packing material itself (Vg)**. This gives us **VT = V0 + Vi + Vg**. To find V0 and Vi, we experimentally measure the **elution volumes (Ve)** of large and small solutes. **Ve is the volume of solvent required to elute a solute from the column after it contacts the gel.**

The elution volume, Ve, of a solute partially included in the pores of the gel bead can be related to V0 and Vi by the equation: **Ve = V0 + σVi**, where **σ is** the partition coefficient of the solute. **The partition coefficient, σ, describes how much of the internal volume is accessible to the solute (0 < σ < 1)**. By comparing σ with values measured for solutes of known size, we can obtain information about the molecular size of an unknown solute. When subjecting a series of solutes with known sizes to size exclusion chromatography, we observe a linear relationship between the partition coefficient and size.

Virtual learning aid

Materials

1. SEC column (packed with porous beads)
2. Sample to be analyzed
3. Mobile phase (buffer)
4. Detector (e.g., UV-Vis)

Protocol

1. **Column preparation**: Choose a size exclusion column with appropriate pore size for your sample. Pack the column with the stationary phase (porous beads).

2. **Sample loading**: Prepare the sample by dissolving it in a suitable buffer solution. Load the sample onto the top of the column.

3. **Elution**: Allow the buffer to flow through the column, carrying the sample components. Larger molecules will elute faster, as they cannot penetrate the pores in the stationary phase.

4. **Detection**: Monitor the elution using a detector (e.g., UV-Vis) to observe peaks corresponding to different-sized molecules.

5. **Analysis**: Analyze the chromatogram to determine the molecular weights and sizes of the separated components.

Ion exchange chromatography

Ion exchange chromatography is a good method for separating charged molecules **based on their reversible electrostatic interactions with a charged stationary phase**. The **stationary phase is typically composed** an insoluble matrix with covalently attached ions **(ion exchangers)**. Depending on the experimental conditions, solutes can be negatively charged, positively charged, or neutral. In this technique, **solutes with an opposite charge to that of the ion exchanger in the mobile liquid phase form reversible bonds through electrostatic interactions**.

The strength of these interactions is **influenced by the charge size and charge density of the solute**. Greater charge or charge density leads to stronger interactions. On the other hand, neutral solutes show a minimal affinity for the stationary phase and pass through the column with the eluting buffer. To release the bound solutes, the column is eluted with a buffer of higher ionic strength or pH. **Increasing the buffer's ionic strength displaces the bound solutes** while raising the buffer pH reduces the interaction strength by reducing the solute or resin charge. This technique offers a versatile approach to separate charged compounds effectively.

Ion exchangers consist of an insoluble matrix with chemically bonded charged groups on its surface. These charged groups determine the type of ion exchanger: **cationic or anionic, based on whether they exchange cations or anions. Cation exchangers**, also known as acidic ion exchangers, are employed for **cation separation**, while **anion exchangers**, also called basic ion exchangers, are used for **anion separation**.

Both cation and anion exchangers can further be classified as **strong or weak depending on the ionizing strength of their functional groups**. For instance, an exchanger with a quaternary amino group would be considered a strongly basic anion exchanger, while an exchanger with primary or secondary aromatic or aliphatic amino groups would be classified as a weakly basic anion exchanger. On the other hand, a strongly acidic cation exchanger contains the sulfonic acid group. These properties determine the effectiveness and specificity of ion exchange processes for various applications in scientific and industrial settings.

List of ion exchangers commonly used in ion exchange chromatography.

Name	Type	Functional group
Anion exchanger	DEAE-cellulose	Weakly basic
Anion exchanger	QAE-Sephadex	Strongly basic
Anion exchanger	Q-Sepharose	Strongly basic
Cation exchanger	CM-cellulose	Weakly acidic
Cation exchanger	SP-Sepharose	Strongly acidic

| **Cation exchanger** | SOURCE S | Strongly acidic |

Virtual learning aid

An ion exchanger is a solid matrix containing either **positively charged groups (cation exchanger) or negatively charged groups (anion exchanger) covalently attached to its surface**. When a mixture of charged solutes is passed through the column, solutes with charges opposite to that of the ion exchanger will bind to it, while others won't. By adjusting the pH or increasing the salt concentration of the eluting buffer, the bound molecules can be released. This process allows for the elution of bound molecules based on their net positive charge density.

Materials

1. Ion exchange column (cation or anion exchange)
2. Sample to be analyzed
3. Buffers with different pH values
4. Detector (e.g., UV-Vis

Protocol

1. **Column preparation**: Choose an ion exchange column (cation or anion exchange) **based on the charge of the analytes**. Pack the column with the appropriate stationary phase.

2. **Sample loading**: Prepare the sample by dissolving it in a buffer compatible with the stationary phase. Load the sample onto the top of the column.

3. **Washing**: Use a buffer with low ionic strength **to remove unbound molecules**.

4. **Elution**: Increase the ionic strength of the buffer to **elute the bound molecules** based on their affinity for the stationary phase.

5. **Detection**: Monitor the elution using a detector (e.g., UV-Vis) to observe peaks corresponding to separated analytes.

6. **Analysis**: **Analyze the chromatogram** to identify and quantify the separated components based on their charge and affinity for the stationary phase.

Selection of Ion exchanger

Before selecting an appropriate ion exchanger for the purification of a biomolecule, one must consider the **nature of the solutes to be separated**. The decision to use a cationic or anionic exchanger **depends largely on the biomolecule's isoelectric point (pI)**. When a solute molecule carries only one type of charged group, the choice is straightforward: a positively charged solute will bind to a cationic exchanger, and a negatively charged solute will bind to an anionic exchanger. However, some solutes have multiple ionizing groups and may possess both positive and negative charges. The overall charge of such molecules is pH-dependent. At the isoelectric point, the solute bears no net charge and won't bind to any ion exchanger. Above the pI, the solute becomes negatively charged and adsorbs to an anion exchanger, whereas below the pI, the solute acquires a net positive charge and binds to a cation exchanger. The pH value, therefore, plays a critical role in determining the appropriate ion exchange for effective purification.

1. **pH > pI**, net negative charge, **binds to the anion exchanger**
2. **pH = pI**, no net charge, **no binding to the ion exchanger**
3. **pH < pI**, net positive charge, **binds to the cation exchanger**

In ion exchange chromatography, the selection of the appropriate ion exchanger for purifying biomolecules depends on their charge characteristics and stability over a range of pH values. Biomolecules with both positive and negatively charged groups should ideally bind to both anionic and cationic exchangers. However, for larger biomolecules, it is crucial to consider the pH range of stability, which refers to the pH range where the biomolecule remains structurally intact.

For instance, if the isoelectric point (pI) of a protein is 4 (meaning it is neutral at pH 4), it is recommended to choose an anion exchanger that binds to the protein at a pH higher than 4 when the protein carries a negative charge. An alternative option would be to use a cation exchanger at a pH lower than 4, but this may not be suitable for many proteins due to stability issues or aggregation. Similarly, if the protein's pI is 10 (positively charged at pH 7), a cation exchanger, which carries a negative charge at neutral pH, should be preferred.

Often, the exact pI of the protein is not known, requiring a trial-and-error approach. Small samples of the mixture are tested with different ion exchangers, and the one resulting in a relatively low level of added solute in the supernatant is chosen for purification. This method can also be extended to determine the conditions for eluting the desired macromolecule from the ion exchanger by subjecting it to buffers with increasing ionic strength or altering pH and analyzing the released macromolecule in each case.

Choice of buffer

The selection of an appropriate buffer system for ion exchange chromatography is crucial and involves considering several factors. These factors include the choice of **buffer substance, pH level**, and **ionic strength**. Buffer ions can interact with ion exchange resins, and those with opposite charges to the ion exchanger can compete with the target solute for binding sites, leading to a significant reduction in column capacity. To optimize the separation process, **cationic buffers should be used with anionic exchangers**, and **anionic buffers should be used with cationic exchangers**. The pH of the buffer should be chosen based on the stability range of the macromolecule to be separated and should also facilitate the binding of the desired macromolecule to the ion exchanger. Additionally, maintaining a relatively low ionic strength is essential to prevent the damping of interactions between the solute and the ion exchanger. **It is recommended to use buffer concentrations in the range from 0.05 to 0.1 M.**

Affinity chromatography

Affinity chromatography is a powerful technique used for purifying biomolecules **based on their specific biological functions or individual chemical structures**. In this method, a ligand (binding substance) is covalently attached to a solid chromatographic matrix. The substance to be purified is applied to the column, and under suitable conditions, it selectively binds to the immobilized ligand. This results in the **retention of the target biomolecule** while other unwanted substances are washed away.

To recover the purified biomolecule, experimental conditions are adjusted to promote its release from the ligand. The process involves several key steps: **selecting an appropriate ligand, immobilizing the ligand** on a support matrix, **binding the molecules of interest to the ligand, removing non-specifically bound molecules**, and finally **eluting the purified biomolecules**. Affinity chromatography is widely used in biotechnology and biochemistry for efficient and specific purification of various biomolecules, making it an essential tool in modern research and industrial applications.

List of Typical biological interactions used in affinity chromatography.

Type of Ligand	Target Molecules or Molecules of Interest
Enzyme	Substrate analog, inhibitor, cofactor

Antibody	Antigen
Lectin	Polysaccharide, glycoprotein, cell surface receptor
Nucleic acid	Complementary base sequence, nucleic acid binding proteins
Avidin	Biotin
Calmodulin	Calmodulin-binding molecule
Poly(A)	RNA containing poly(U) sequences
Glutathione	Glutathione-S-transferase or GST fusion proteins
Proteins A and G	Immunoglobulins

Virtual learning aid

For example, Eukaryotic mRNA containing a poly(A) tail can be isolated from other RNA molecules using oligo(dT)-cellulose affinity chromatography. The poly(A) tail interacts strongly with short oligo(dT) chains immobilized on a solid support. High-salt conditions stabilize the formed dT-A base pairs, enabling selective binding of poly(A) RNA. Subsequent use of a low-salt buffer disrupts the double-stranded structures, facilitating elution of the poly(A) RNAs from the resin.

DNA affinity chromatography

DNA affinity chromatography is a technique used to **purify specific DNA-binding proteins**. It involves attaching a chemically synthesized double-stranded oligonucleotide with the desired DNA sequence to an insoluble matrix like agarose, creating a column that selectively captures proteins recognizing that specific DNA sequence.

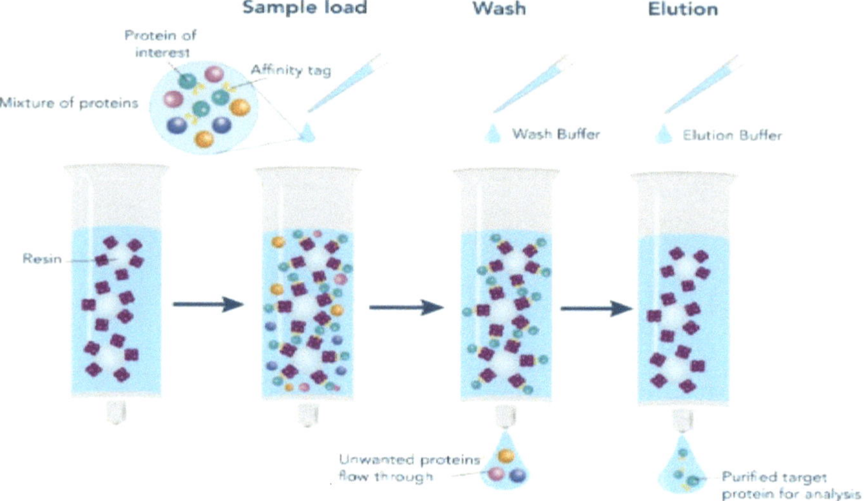

Materials

1. Affinity column with immobilized DNA-binding proteins or ligands
2. Sample containing DNA
3. Buffers
4. Detector (e.g., UV-Vis)

Protocol

1. **Column preparation**: Prepare an affinity column by immobilizing DNA-binding proteins or ligands on the stationary phase.

2. **Sample loading**: Prepare the sample containing DNA in a suitable buffer and load it onto the column.

3. **Washing**: Wash the column with a buffer to remove unbound molecules.

4. **Elution**: Elute the bound DNA by using a buffer that disrupts the protein-DNA or ligand-DNA interactions.

5. **Detection**: Monitor the elution using a detector (e.g., UV-Vis) to observe peaks corresponding to the DNA fragments.

6. **Analysis**: Analyze the chromatogram to identify and quantify the DNA fragments based on their interactions with the immobilized proteins or ligands.

Chapter 6

Spectroscopy

Introduction

Spectroscopy is the fascinating **study of how matter interacts with electromagnetic radiation based on its wavelength or frequency**. Matter, which can be atoms, molecules, or ions, engages in various ways with radiation, such as absorption, emission, or scattering. These interactions allow us **to quantitatively or qualitatively investigate both the matter and physical processes**. By analyzing the radiation absorbed or emitted by an atom or molecule, we can learn about its identity (**qualitative spectroscopy**), while measuring the total radiation provides insights into the number of absorbing or emitting atoms or molecules (**quantitative spectroscopy**).

Electromagnetic radiation

Electromagnetic radiation is a type of energy that combines electrical and magnetic properties. It travels as waves with the electric and magnetic fields oscillating at right angles to the direction of propagation. The electromagnetic spectrum covers a wide range, from very short wavelengths like gamma rays to much longer ones like radio waves. In the visible region, wavelengths between 400 and 700 nanometers create the colors we see, with shorter wavelengths appearing blue and longer ones appearing red. There's also the near ultraviolet region (200-400 nm) and the infrared region (above 700 nm to around 2000 nm). These divisions are not strict, and some overlap between regions can occur.

Electromagnetic radiation possesses essential characteristics like speed, amplitude, frequency, energy, and polarization. Energy in a specific part of the spectrum is linked to the frequency and wavelength of the radiation. **Frequency (ν)** denotes the **number of wave cycles passing a point in one second**, measured in **Hz (1 Hz = 1 cycle/sec). Wavelength (λ)** represents the **length of a complete wave cycle and is usually measured in centimeters**. The relationship between frequency and wavelength is governed by $\nu = c/\lambda$, where **c is the speed of light**.

The energy of electromagnetic radiation is directly proportional to its frequency and inversely proportional to its wavelength. This relationship is defined by the equation $E = h\nu = hc/\lambda$, **where h is Planck's constant (6.6x10^-34 joules-sec)**. In the **infrared region** of the electromagnetic spectrum, **wavenumber (V) is utilized instead of wavelength. Wavenumber is simply the reciprocal of wavelength governed by $V = 1/\lambda$** and is expressed in **units of per centimeter (cm^{-1})**. Wavenumbers are favored by chemists because they are directly associated with energy—the higher the wavenumber, the higher the energy.

Virtual learning aid

Types of spectroscopy

Spectroscopy is a group of techniques that **study how electromagnetic radiation interacts with the substance** we want to examine. When the radiation meets the matter, it can be **reflected, scattered, emitted, transmitted, or absorbed.** This leads to **three main types of spectroscopy: absorption, emission,** and **scattering spectroscopy.**

Absorption spectroscopy

Absorption spectroscopy is a technique that **studies how a sample absorbs different wavelengths of electromagnetic radiation.** When the radiation passes through the sample, some wavelengths are absorbed, causing a decrease in intensity. These absorbed wavelengths create a unique pattern called an **absorption spectrum.** By analyzing this spectrum, we can learn valuable information about the sample's composition. The absorption process occurs when the energy of the radiation matches the energy difference between two states of the atom or molecule. Different types of absorption spectroscopy, like X-ray, UV-Vis, and infrared spectroscopy, focus on specific regions of the electromagnetic spectrum for analysis.

Note

Mass spectrometry and spectroscopy are distinct techniques; unlike spectroscopy, which examines the absorption or emission of radiation, mass spectrometry focuses on determining the molecular mass.

List of different types of spectroscopy.

Electromagnetic Spectrum	Spectroscopic Type
X-ray	X-ray spectroscopy
UV-Vis	UV-Vis spectroscopy
Infrared (IR)	Infrared spectroscopy, Raman spectroscopy
Microwave	Microwave spectroscopy
Radio wave	Electron spin resonance spectroscopy, Nuclear magnetic resonance spectroscopy

Effect of interaction between electromagnetic radiation and matter

Atomic and molecular spectroscopies deal with how substances interact with electromagnetic radiation, like absorbing, emitting, or scattering it, leading to changes in the systems. In absorption spectroscopy, the impact of the radiation depends on its energy. Very energetic radiations (like UV and X-rays) can even eject electrons (ionization), whereas infrared radiation has less energy, affecting the vibrational energy of chemical bonds in molecules. Even less energetic, microwave radiation can't cause electronic transitions or vibrations but only makes molecules rotate.

Region of Spectrum	Types of Energy Transitions
X-rays	Ionization
UV Radiations	Ionization and Electronic Transition
Visible Radiations	Electronic Transition
Infrared	Molecular Vibration
Microwaves	Molecular Rotation
Radio Waves	Nuclear Spin (in the case of Nuclear Magnetic Resonance)
	Electronic Spin (in case of Electron Spin Resonance)

Spectrophotometer

In analytical chemistry, light absorption plays a vital role in characterizing and quantifying substances. To accomplish this, we use a powerful instrument called a **spectrophotometer**, which **measures the amount of light transmitted through a sample at a specific wavelength**. It compares the intensity of light before and after passing through the sample, enabling us to determine the quantity of light absorbed. Essentially, a spectrophotometer is a **combination of a spectrometer and a photometer**, utilizing a monochromator to select the wavelength for measurement. The **key components** of a spectrophotometer include a **light source**, an **optical system or monochromator**, a **sample holder (usually a cuvette)**, and a **light detector**.

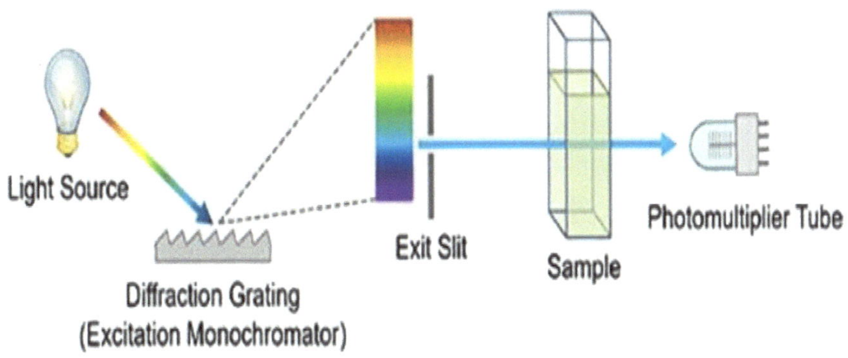

Note

1. Monochromatic: It refers to electromagnetic radiation that consists of only one single wavelength, like a single color of light.

2. Polychromatic: This term describes electromagnetic radiation that contains multiple wavelengths, like a combination of colors of light.

3. Photometer: An instrument used to measure the intensity of a beam of light, providing information about its brightness.

4. Spectrometer: An apparatus designed for recording and measuring spectra, which are patterns of light that reveal information about the composition of a sample.

Fundamental concept of Absorption spectroscopy

When light passes through a material, some of it gets absorbed, leaving behind a spectrum with a gap called an **absorption spectrum** when the remaining light is passed through a prism. This unique absorption spectrum is specific to a particular element or compound and remains constant, regardless of its concentration.

The Beer-Lambert law establishes that at a given wavelength, the amount of absorbed light is directly proportional to the concentration of the absorbing substance and the sample's thickness.

Beer-Lambert law

In simple terms, when light shines on a substance, three things can happen: some light bounces back (reflects), some light gets absorbed by the material, and the rest passes through it. The process of light absorption is described by two laws - **Lambert's law** and **Beer's law**, which together form the Beer-Lambert law.

1. **Lambert's law** tells us that when a specific kind of light passes through a see-through substance, the **intensity of the light** that comes out on the other side **decreases as the material becomes thicker**.

2. **Beer's law** states that the **intensity of light** that comes through a substance **decreases as the concentration of the substance increases**.

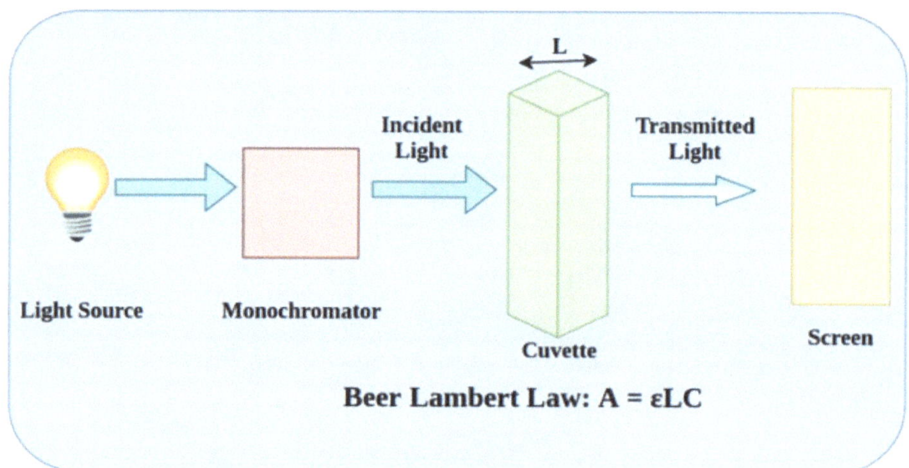

Mathematical expression of Beer-Lambert law

The relationship between concentration, length of the light path, and the light absorbed by a particular substance is expressed mathematically as shown below:

$A = \log I_o / I = \varepsilon.c.l$

where,

A = Absorbance

I_o = Intensity of incident light

I = Intensity of light transmitted through the sample

ε = Extinction coefficient or absorption coefficient for an absorbing compound

c = **Concentration of absorbing material** in the sample
l = **Path length (cm)**

If the concentration is expressed in molarity, ε **is termed the molar absorption coefficient or molar extinction coefficient**. Its unit is $\mathbf{M^{-1}\ cm^{-1}}$. If the concentration is expressed in g/liter, ε **becomes the specific absorption coefficient**. The **absorbance is a dimensionless** quantity. Theoretically, absorbance (A) can have any positive value; in practice, for UV and visible spectrometers, **'A' normally varies between zero and one**.

The ratio of the intensity of the transmitted light (I) to the intensity of the incident light (Io) is called transmittance (T).

$$T = I/Io$$

It measures the amount of light transmitted after passing through the medium. The smaller the transmittance, the greater the absorption of light. Transmittance is always a numerical value between zero (all light absorbed) and one (no light absorbed). It is common to convert transmittance into percent transmittance (%T). It is $100 \times T$.

$$\%T = I/Io \times 100\% = T \times 100\%$$

Percent transmittance varies between 100% (all light transmitted) and 0% (no light transmitted- ted). The transmittance and absorbance are inversely related. Moreover, the inverse relationship between transmittance and absorbance is not linear, it is logarithmic.

$$A = \log_{10} 1/T$$

Virtual learning aid

UV/VIS absorption spectroscopy

UV/VIS spectroscopy is a technique that analyzes the absorption of ultraviolet (UV) and visible (VIS) light by a sample. It provides valuable information **about the electronic structure and concentration of analytes in a solution.**

Fundamental concept

When transparent materials are exposed to electromagnetic radiation, they can absorb some of the radiation, causing their **atoms or molecules to transition from a lower energy state (ground state) to a higher energy state (excited state)**. The absorbed radiation possesses precisely the energy needed for this transition, matching the difference in energy between the ground and excited states.

Virtual learning aid

Note

1. During an electronic transition, an electron undergoes a shift from one orbital to another. This transition can take place within atoms, involving atomic orbitals, or within molecules, involving molecular orbitals.

2. Ultraviolet and visible absorption spectroscopy relies on electron transitions between molecular orbitals caused by absorbing electromagnetic radiation in the UV and visible range. When a molecule absorbs energy, an electron moves from an occupied orbital to an unoccupied orbital with higher potential energy. Typically, this transition occurs from the highest occupied molecular orbital (HOMO) to the lowest unoccupied molecular orbital (LUMO).

3. Atomic spectroscopy refers to the study of the electromagnetic radiation absorbed and emitted by atoms whereas molecular spectroscopy refers to the study of the electromagnetic radiation absorbed and emitted by molecules.

When light passes through a compound, it can excite electrons from their normal positions in **bonding or non-bonding orbitals to higher energy levels in empty anti-bonding orbitals**. The energy gaps between these levels determine which wavelengths of light the compound can absorb. In the UV and visible region, **molecules with unsaturated functional groups containing π bonds or atoms with non-bonding orbitals are capable of absorbing light**. These molecules are known as **chromophores**, meaning **"to bear color"** in Greek. Molecules with conjugated systems, like alternating single and double bonds, show even lower energy gaps between orbitals, allowing them to absorb longer wavelengths of light. As the delocalization of electrons increases in the molecule's structure, the energy gap decreases, resulting in absorption of light with lower energy and longer wavelengths.

Transition	Wavelength (nm)
σ to σ*	< 200
n to σ*	160-260
π to π*	200-500
n to π*	250-600

Materials and Equipment

1. UV/VIS spectrophotometer
2. Quartz cuvettes with appropriate path length (e.g., 1 cm)
3. Sample solutions in suitable solvents
4. Blank solution (solvent without analyte)
5. Distilled water
6. Pipettes and micropipettes with appropriate volumes
7. UV/VIS compatible containers or vials for sample preparation
8. Kimwipes or lint-free tissues for cleaning cuvettes

Protocol

1. **Instrument Preparation**: Turn on the UV/VIS spectrophotometer and allow it to warm up for the recommended time specified by the manufacturer. Check that the instrument is properly calibrated. **Perform a wavelength calibration using certified reference materials, if required**.

2. **Sample Preparation**: Prepare your sample solutions with the appropriate concentration. Dilute the stock solution using a suitable solvent to obtain a concentration within the linear range of the instrument. **Label your cuvettes and the corresponding solutions to avoid mix-ups. Always include a blank**

solution containing the solvent without the analyte to account for background absorbance.

3. **Blank Measurement**: Place the blank cuvette in the sample holder. Set the wavelength range according to your experiment's requirements. Adjust the baseline by zeroing the instrument using the blank solution.

4. **Sample Measurement**: Remove the blank cuvette and carefully insert the sample cuvette in the sample holder. Set the desired wavelength for the measurement. Record the absorbance value, ensuring the cuvette is correctly aligned and free from air bubbles or scratches.

5. **Data Analysis**: Calculate the absorbance of each sample by subtracting the blank absorbance value from the sample absorbance value. Prepare a calibration curve by measuring absorbance values at different concentrations of the analyte. Determine the concentration of the unknown sample using the calibration curve.

6. **Cleaning and Maintenance**: After completing the measurements, remove the cuvettes and rinse them thoroughly with distilled water, followed by the solvent used in the analysis. Clean the exterior of the spectrophotometer with a lint-free tissue.

IR absorption spectroscopy

Infrared (IR) radiation spans the electromagnetic spectrum from **1 micrometer to 100 micrometers**, encompassing **near-IR, mid-IR**, and **far-IR** sub-regions. Each IR frequency carries specific energy levels, and when it passes through a substance under investigation, its energy is transferred to the compound. Molecules absorb only certain IR frequencies, leading to excitation to higher energy states, with energy changes typically ranging from 8 to 40 kJ/mole. Unlike UV and visible radiations, IR lacks the energy to induce electronic transitions, focusing instead on bond vibrations due to their corresponding energy levels.

Fundamental concept

Infrared (IR) spectroscopy, a type of **vibrational spectroscopy**, examines the vibrations of molecular bonds at temperatures above absolute zero. These vibrations primarily involve **two types: stretching, which occurs along the chemical bond,** and **bending, which occurs away from the bond line**. Specific bonds vibrate at characteristic frequencies, like the C-H bonds in organic compounds. When an electromagnetic wave matches the frequency of a particular C-H stretching vibration (e.g., 9×10^{13} times per second), energy is absorbed, resulting in an increased frequency and intensity of the bond vibration. This absorption manifests as peaks in the IR spectrum, and eventually, the bond returns to its normal vibration frequency, releasing energy in the form of heat.

Different bonds exhibit distinct vibrational frequencies, leading to varied IR absorption frequencies. When IR radiation is absorbed, both the frequency and intensity of the bond vibration amplify, signifying a higher frequency and larger amplitude of the vibrational motion. By identifying these absorption patterns, IR spectroscopy helps characterize molecular structures and compositions.

An **IR-active** normal mode of vibration in the infrared spectral range generates an **oscillating dipole**, allowing for IR absorption. On the other hand, an **IR-inactive** mode, like the stretching of a **homonuclear diatomic molecule, lacks an oscillating dipole moment** and does not lead to IR absorption. The dipole moment is determined by how electrons are distributed along the bond and its length. When bonding occurs between different atoms, a **higher difference in electronegativity leads to a greater dipole moment**.

Infrared (IR) radiation can be absorbed by certain chemical bonds in a molecule, but not all bonds are capable of this. **For IR absorption** to occur, the molecule must possess a **changing dipole moment** during vibration and have a **polar chemical bond with a bond dipole**. The **magnitude of the bond dipole** depends on the **charge on each bonded atom** and **the bond length**. When the frequency of IR radiation matches that of the vibrating bond dipole, energy is transferred, making the vibration infrared-active. However, vibrations that do not lead to IR absorption are termed infrared-inactive.

Diatomic molecules like H2, N, and O, which have two identical atoms, **lack a dipole moment** and are therefore **infrared-inactive**. Diatomic molecules with different atoms usually have a permanent dipole moment due to differences in electronegativity. However, having a permanent dipole moment does not automatically make a molecule infrared-active. For a vibration to be infrared-active, it must cause a change in the dipole moment of the molecule. Even a molecule with a zero-dipole moment can become infrared-active if a specific molecular vibration creates a temporary dipole moment, as seen in the example of CO2. The CO2 molecule is symmetrical and has a zero equilibrium dipole moment. Symmetric stretching vibrations, where both bonds are compressed or stretched equally, do not cause a change in dipole moment and are, therefore, infrared-inactive. However, in asymmetric stretching, one bond compresses while the other stretches, leading to a change in bond length and dipole moment, making it infrared-active. Bending vibrations in CO2 also result in a changing dipole moment, rendering the molecule infrared-active.

Virtual learning aid

Scan It!

Infrared spectra

An **infrared spectrum**, serving as an **absorption record**, showcases the IR radiation absorbed by a compound as it varies with wavelength. To analyze this, we employ an **infrared spectrometer**. This spectrum also reveals the functional groups present in the compound. The **position of the peak** (wave number or wavelength) and **its intensity** are crucial aspects of IR spectra, conveying valuable information about the compound's characteristics.

IR absorption position

The energy associated with bond vibrations in molecules depends on several factors.

Firstly, the **strength of the bond** plays a crucial role. **Stronger bonds**, characterized by **higher bond dissociation energies**, exhibit IR absorptions at **higher wavenumbers** in an infrared spectrum. For instance, triple bonds have higher stretching frequencies compared to double bonds, which, in turn, have higher frequencies than single bonds.

Secondly, the **masses of the atoms** forming the bond also influence the vibration frequency. Bonds involving **lighter atoms**, like hydrogen, tend to have **higher stretching** frequencies than those with heavier atoms.

Additionally, the **type of vibration** being observed affects the absorption frequency. Molecules undergo **two general types** of vibrations: **stretching** and **bending**. **Bending vibrations generally occur at lower frequencies** (higher wavelengths) than stretching vibrations for the same functional groups. This arises from the fact that it is easier to bend a bond than to stretch or compress it. Consequently, different bonds within a molecule vibrate in distinct ways, absorbing varying frequencies (and energies) of infrared radiation.

IR absorption intensity

In an infrared (IR) spectrum, the **peaks represent different molecular vibrations,** and their intensities can vary significantly. The **intensity of an IR absorption** is influenced by **two key factors**. Firstly, it depends on the **concentration of molecules** in the sample; a higher concentration leads to stronger IR absorption. Secondly, the **magnitude of the dipole moment** plays a crucial role—the greater the change in dipole moment during a vibration, the more pronounced the absorption. **Symmetrical bonds**, where there is no change in dipole moment, result in **no IR absorption**. The bond dipole's magnitude is determined by the charge on each bonded atom and the bond length, making it an essential factor in the intensity of IR absorption.

Infrared absorption and chemical structure

Infrared (IR) spectroscopy finds its most common application in **identifying functional groups** within compounds. The distinct vibrations of different functional groups occur **at specific frequencies**, facilitating their identification. Typically, a particular type of functional group absorbs IR radiation within a specific region of the IR spectrum. Nevertheless, the frequency of vibration is **influenced by factors like electron delocalization, hydrogen bonding**, and **substitutions in nearby groups**.

An **IR spectrum** generally spans from approximately **4000 cm^{-1} to 650 cm^{-1}**, and it can be divided into the **functional group region** and the **fingerprint region**. The functional group region, ranging from **4000 cm^{-1} to 1500 cm^{-1}**, allows the identification of functional groups, with certain absorption bands serving as diagnostic markers. For example, C-H stretching vibrations appear between 3200 cm^{-1} and 2800 cm^{-1}, while carbonyl (C=O) stretching vibrations are usually observed between 1800 cm^{-1} and 1600 cm^{-1}.

The **fingerprint region**, spanning from **1200 cm^{-1} to 700 cm^{-1}**, is more complex and challenging to interpret. It contains a variety of absorptions, including those of different vibrations and bonds like C-O, C-C, C-N single bond stretches, C-H bending vibrations, and bands related to benzene rings. Since the natural frequency of vibration varies for each type of bond, and the environment of identical bond types in different compounds differs, no two molecules with different structures exhibit precisely the same infrared absorption pattern or spectrum. Therefore, by comparing the infrared spectra of two substances, their identity or non-identity can be established, and the infrared spectrum can also provide valuable structural information about a molecule.

Property	Position of Band	Strength of Band	Width of Band
Mass of Atoms	Light atoms give high-frequency	-	-
Bond Strength	Strong bonds give high-frequency	-	-
Change in Dipole Moment	-	A large change gives strong absorption	-

Hydrogen Bonding	-	-	Strong H-bond gives wide pea

Materials

1. **IR Spectrometer**: A **Fourier-transform infrared (FTIR)** spectrometer with a suitable light source (e.g., a globar) and a detector (e.g., a liquid nitrogen-cooled mercury cadmium telluride detector).
2. **Sample Holder**: IR-transparent material (e.g., potassium bromide - KBr) for preparing solid samples, and IR-transparent cells (e.g., quartz or potassium bromide) for liquid samples.
3. **Sample Preparation Tools**: Mortar and pestle for grinding solid samples, and a syringe for preparing liquid samples.
4. **Deuterated Solvent**: For analyzing liquid samples in deuterated solvents, e.g., deuterated chloroform (CDCl3).
5. **Tissues/Paper**: To handle samples without contaminating them with fingerprints or other materials.
6. **Nitrile Gloves**: To avoid contamination and protect the operator.

Protocol

1. **Preparation**: Ensure the IR spectrometer is powered on and calibrated according to the manufacturer's guidelines. Check that the sample compartment is clean and free of debris. Load the appropriate IR-transparent material (KBr or quartz) into the sample holder, ensuring there are no visible scratches or impurities.

2. **Sample Preparation**:
A. Solid Samples: Grind the solid sample with a mortar and pestle until it becomes a fine powder. Mix the powdered sample with dry KBr in a suitable ratio (typically 1:100 to 1:10, w/w) and press the mixture into a pellet using a hydraulic press. Place the pellet in the sample holder.
B. Liquid Samples: Fill a clean IR-transparent cell (quartz or KBr) with the liquid sample, ensuring it covers the path of the IR beam. Carefully insert the cell into the sample holder.

Note

3. **Spectral Acquisition**: Open the software for the IR spectrometer and select the appropriate measurement mode (transmission or attenuated total reflectance - ATR). Align the sample holder in the spectrometer, ensuring the sample is in the path of the IR beam. Set the desired wavelength range and resolution for the analysis. Lower resolution for broad features and higher resolution for fine details. Initiate the spectral acquisition, ensuring the system collects an adequate number of scans to improve signal-to-noise ratio.

4. **Data Analysis**: Once the spectral acquisition is complete, the software will display the IR spectrum. Identify and analyze characteristic peaks and bands in the spectrum. Compare them to reference spectra or databases to determine the functional groups present in the sample.

Nuclear Magnetic Resonance

Nuclear Magnetic Resonance (NMR) is a powerful spectroscopic technique used **to study the magnetic properties of atomic nuclei**. It involves **detecting changes in nuclear spin energy** when exposed to an external magnetic field. Nuclei have a property called nuclear spin, which gives rise to their magnetic behavior.

Fundamental concept

NMR allows us **to identify the type of environment in which atomic nuclei exist within a molecule**. For instance, in a molecule like propanol, the hydrogens in the hydroxyl group have a different environment compared to those in the carbon skeleton. By using proton NMR (also known as H NMR or proton NMR), we can easily distinguish between these different types hydrogens. Similarly, carbon NMR (C NMR) enables the distinction between the three different carbon atoms in propanol.

However, not all atomic nuclei have nuclear spin. **Only nuclei with a magnetic property due to nuclear spin can be detected using NMR**. The concept of nuclear spin is quantized and can have multiples of $1/2$, with the spin being either positive (+) or negative (-). The net spin of a nucleus depends on the number 5of protons and neutrons it possesses:

1. If both the **number of neutrons and protons are even**, the individual spins cancel out, resulting in **zero overall spin** (nuclei with no spin).

2. If both the **number of neutrons and protons are odd**, the nucleus has an **integer spin** (1, 2, 3, etc.).

3. If the **sum of the number of neutrons and protons is odd**, the nucleus has a **half-integer spin** (1/2, 3/2, 5/2, etc.).

Element	Number of Protons (Z)	Number of Neutrons (N)	Nuclear Spin Quantum Number (I)
Hydrogen	1	0	1/2
Helium	2	2	0
Carbon	6	6	0
Nitrogen	7	7	1
Oxygen	8	8	0
Sodium	11	12	3/2
Iron	26	30	0
Copper	29	35	3/2
Silver	47	61	5/2
Gold	79	118	1/2

Note

In the absence of an external magnetic field, nuclear magnetic poles are randomly oriented. However, when exposed to an external magnetic field, nuclei with a spin state of +1/2 align parallel to the field, while those with a spin state of -1/2 align antiparallel to it. This alignment along the applied field occurs because it represents a lower energy configuration.

Virtual learning aid

Nuclear Magnetic Resonance (NMR) spectroscopy is founded on the magnetic properties of atomic nuclei **arising from a phenomenon called**

nuclear spin. Just like tiny magnets, nuclei with spin possess distinct energy states. This analogy can be likened to a magnetic compass in an external magnetic field. Imagine turning off the Earth's magnetic field—the compass needle would point randomly. When the magnetic field is reinstated, the needle aligns north, its lowest energy state. For atomic nuclei, placed in an external magnetic field, they exhibit different energy levels, and the number of energy levels depends on the nucleus type, such as 1H or C13, with hydrogen nuclei referred to as "protons."

Let's focus on the hydrogen nuclei within a chemical sample. In absence of an external magnetic field, the nuclear magnetic poles are randomly oriented. However, when an external magnetic field is applied, the 1H nuclei can either align parallel, their lowest energy state, or antiparallel, higher in energy, to the field. Nuclei with a spin of +1/2 orient parallel to the field, while those with a spin of -1/2 orient antiparallel to the applied field. This creates two distinct energy levels for 1H nuclei in the magnetic field.

Unlike a compass needle, which can rotate freely through 360° with countless energy levels, atomic nuclei acting as tiny magnets exhibit specific energy level characteristics when placed in an external magnetic field.

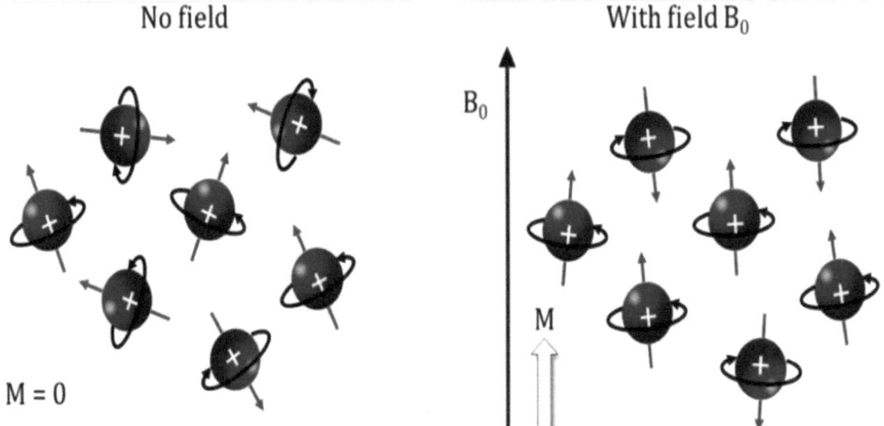

In the absence of an external magnetic field, the two spin states possess equal energy. Yet, when an external magnetic field is introduced, the two spin states acquire different energies: the +1/2 spin state holds lower energy compared to the -1/2 spin state. This energy difference depends on the strength of the applied magnetic field. The stronger the magnetic field, the greater the energy difference between the possible spin states. However, the energy difference between the two nuclear spin states is exceptionally small - necessitating an extremely powerful magnetic field to observe any discernible distinction.

In an external magnetic field, a nucleus experiences a difference in energy between its two spin states. This energy gap depends on the strength of the magnetic field and the specific magnetic properties of the nucleus itself. We describe this phenomenon using the fundamental equation of Nuclear Magnetic Resonance (NMR): $\Delta E = \gamma \hbar B$

Here's what each component means:

1. ΔE represents the **energy difference between the two spin states**, measured in joules or electron volts.

2. γ denotes the **gyromagnetic ratio**, a constant unique to the nucleus under study, measured in radians per second per Tesla.

3. $\hbar$ stands for the **reduced Planck constant**.

4. B_0 indicates the **strength of the static magnetic field**, measured in Tesla.

When there's **no external magnetic field ($B_0 = 0$)**, there's **no energy difference between the spin states**. As we increase the magnetic field, the energy gap between the two spin states also increases. In a chemical sample containing 1H nuclei, each nucleus occupies one of two spin states, which differ in energy by ΔE. Most 1H nuclei prefer the spin state with +2, but there's a small excess in the spin state with -2.

Even under a very strong magnetic field, the population difference between the two spin energy states remains very small. However, if we subject the sample to electromagnetic radiation with energy exactly matching ΔE, some 1H nuclei in the +2 spin state will absorb this energy. Consequently, these protons "flip" their spins and move to a more energetic state with a spin of -2. Since the energy difference between the two states is tiny even in a strong magnetic field, electromagnetic radiation of radio wave frequency suffices to flip the nuclei. Radio waves can cause the nucleus to transition from a lower energy state to the higher one.

This fascinating absorption phenomenon, known as nuclear magnetic resonance, can be detected using a specialized type of absorption spectrometer called an **NMR spectrometer**. It enables scientists to gain valuable insights into the magnetic properties and structures of molecules in a sample. In nuclear magnetic resonance (NMR) spectroscopy, the **chemical shift refers to the different absorption frequencies of protons** (hydrogen nuclei) in a molecule.

Note

The proton NMR (H NMR) spectrum is a graphical representation of energy absorption versus the relative frequency of the absorbed radiation. The peaks in this spectrum, referred to as absorptions or lines, are positioned according to their chemical shift, typically expressed in parts per million (ppm). Notably, the small peak on the far right corresponds to the protons of tetramethylsilane (TMS), a commonly used reference standard in NMR spectroscopy

In an atom, the local diamagnetic currents create an induced magnetic field that counters the external magnetic field's direction, resulting in a reduced effective or local magnetic field (B) sensed by a proton. This phenomenon is known as shielding, and it protects the nucleus from the external magnetic field. The extent of shielding depends on the electron density surrounding the nucleus: greater electron density enhances shielding, while lower density diminishes it.

Atoms with relatively low electronegativity neighboring a proton boost electron density around that proton, enhancing its shielding from the applied magnetic field. Conversely, atoms with higher electronegativity decrease the electron density around a proton, leading to reduced shielding. In summary, the more-shielded protons experience a smaller local magnetic field compared to less-shielded protons, owing to the influence of neighboring electron densities. This phenomenon of shielding is fundamental to understanding the behavior on protons in molecules and is critical in various scientific contexts.

Note

1. Changes in the electron distribution around a nucleus have significant effects: altering the local magnetic field experienced by the nucleus, influencing the resonance frequency of the nucleus, and impacting the chemistry of the molecule at that specific atom.

2. The chemical shift is a measure of resonance frequency compared to a reference compound set at 0 ppm, expressed in parts per million (ppm), independent of the spectrometer frequency, making it more convenient for analysis.

In NMR spectroscopy, unlike IR and UV/Vis spectroscopy, the signals are dependent on the field strength, making it necessary to specify the signals location. To tackle this, we use **TMS as a reference sample, assigning it a chemical shift of 0 ppm**. The chemical shift of other nuclei relative to TMS indicates their absorption frequencies. This scale standardizes the measurements regardless of the spectrometer's magnetic field strength. By employing TMS as a reference, NMR spectra become more consistent and reliable for chemical analysis.

Materials

1. **NMR Instrument**: A high-field NMR spectrometer, preferably operating at 300 MHz or higher, equipped with a suitable probe for the sample type.
2. **NMR Tubes**: High-quality NMR sample tubes made of glass with precise dimensions (typically 5 mm in diameter) and clean inner walls.
3. **NMR Solvents**: Deuterated solvents such as CDCl3, D2O, DMSO-d6, etc., are appropriate for dissolving the sample and providing a solvent lock signal.
4. **Sample**: The compound of interest, is purified and dissolved in an appropriate deuterated solvent. Concentration should be optimized to achieve a good signal-to-noise ratio (SNR).
5. **NMR Standard**: Tetramethylsilane (TMS) or any other suitable standard compound for referencing the chemical shifts of other signals.
6. **Syringes or Pipettes**: To measure and transfer the sample and NMR solvents accurately into the NMR tubes.
7. **Sample Storage**: Sample vials and airtight containers for sample storage.
8. **NMR Software**: Software for data acquisition, processing, and analyse NMR spectra.

Protocol

1. **Sample Preparation**: Weigh the purified compound accurately. Dissolve the compound in an appropriate deuterated solvent, ensuring the solution is homogeneous. Transfer the sample solution into a clean NMR tube, ensuring there are no air bubbles.

2. **NMR Instrument Preparation**: Power on the NMR instrument and let it stabilize for a few minutes. Check that the NMR probes and system are clean and free from any contaminants. Load the NMR tube containing the sample into the NMR spectrometer.

3. **NMR Instrument Setup**: Set the temperature to the desired value, if applicable. Tune and match the NMR probe to optimize the instrument's performance. Calibrate the NMR spectrometer using a standard reference like TMS to set the 0 ppm point for the chemical shifts.

4. **NMR Data Acquisition**: Initiate the NMR data acquisition software. Set the desired NMR experiment (e.g., 1D proton, 2D COSY, 2D HSQC, etc.). Adjust acquisition parameters such as pulse width, relaxation delay, number of scans, and spectral width as per the experiment requirements. Start the data acquisition and allow the instrument to record the NMR spectrum.

5. **Data Processing and Analysis**: Process the acquired NMR data using the NMR software to apply Fourier Transform and phase correction. Calibrate the NMR spectrum using the TMS signal as a reference. Assign chemical shifts to the signals in the spectrum based on known NMR data for the solvent and literature references for the compound. Analyze the spectral patterns to determine the connectivity and structural information of the compound.

Circular Dichroism

Circular Dichroism (CD) spectroscopy is a powerful technique used **to investigate the secondary structure and conformational changes in biomolecules, including proteins, nucleic acids, and peptides**. It relies on the differential absorption of left- and right-circularly polarized light by chiral molecules. This technique provides valuable information about the **secondary structure content, folding/unfolding transitions, ligand binding**, and **protein stability**.

Fundamental concept

Light is an electromagnetic wave that propagates transversely. Its electric and magnetic fields oscillate perpendicular to the direction of wave propagation. Depending on the orientation of the electric field, light can be unpolarized, with the electric field vector oscillating in multiple planes, or polarized, with the electric field vector oscillating in a single plane perpendicular to the propagation direction. Linearly polarized light is a specific type of polarization where the electric field vector remains constant in direction while oscillating in magnitude. This polarization is achieved by passing unpolarized light through a polarizer, which absorbs electric field vectors not aligned with a particular plane.

In **circularly polarized light**, the electric field vector rotates around the propagation axis while maintaining a constant magnitude. This is in contrast to linearly polarized light, where the magnitude oscillates while the direction remains constant. **Circularly polarized light is created by superimposing two plane-polarized light waves of the same wavelength and amplitude but with a phase difference of 90°.** This phase difference causes one wave to be at its peak when the other is crossing the zero line. Circularly polarized light can be **either right-handed or left-handed**, depending on the rotation direction of the electric field vector.

Left-handed circularly polarized light, when observed with propagation towards the observer, features an **electric field vector that rotates clockwise**. By superimposing equal-amplitude left and right circularly polarized light beams, linearly polarized light can be obtained. Therefore, linearly polarized light can be thought of as the superposition of opposite circularly polarized light with equal amplitude and phase. This phenomenon provides insight into the fascinating properties and behavior of light as an electromagnetic wave.

Optical rotatory dispersion (ORD) and **circular dichroism (CD)** are intriguing phenomena arising from the interaction of polarized light with chiral molecules—molecules lacking a plane and center of symmetry. Since most biological molecules exhibit **chirality**, they are **considered optically active**. This optical activity gives rise to a remarkable phenomenon known as optical rotation, wherein the plane of polarization for linearly polarized light undergoes rotation while passing through an optically active medium. The ORD technique precisely measures the ability of optically active compounds to rotate plane-polarized light, revealing valuable insights into their wavelength-dependent optical rotation characteristics.

Virtual learning aid

Note

Chiral molecules are characterized by having one or more chiral centers, where a tetrahedral carbon atom bears four distinct substituents. For a molecule to be chiral, it must lack both a plane of symmetry and a center of symmetry. In experiments involving chiral compounds, unpolarized light from a normal light source is transformed into plane-polarized light by a polarizer. This plane-polarized light then passes through a sample tube containing the chiral compound before being analyzed.

Circular Dichroism is a type of absorption spectroscopy where the sample displays different colors when illuminated with right and left-polarized light. The term **"dichroism"** originates from Greek, meaning **"two colors,"** as the **observed color variations arise due to light absorption**. It occurs when optically active substances absorb slightly different amounts of left and right circularly polarized light. This difference in absorption at a specific wavelength is quantified using the **molar extinction coefficient (ε)**. Optically active samples have **distinct molar extinction coefficients for left (εL) and right**

(εR) circularly polarized light, and the variation in absorbance between the two provides a measure of circular dichroism.

Circular Dichroism (CD) in absorption spectroscopy is characterized by the difference in absorption (ΔA) between left and right circularly polarized light. This difference can be expressed as

$$\Delta A = (\varepsilon L - \varepsilon R) \times c \times l,$$

Where,

1. **εL and εR** are the **molar extinction coefficients** for **left** and **right** circularly polarized light
2. **c** is the **molar concentration**
3. l is the **path length (cm)**.

The CD is proportional to the **ellipticity (θ)** of the transmitted light, and their relationship is given by $\theta = 33 \times \Delta A$ **degrees**.

CD data can be presented in terms of ellipticity (θ) or **differential absorbance (ΔA)**. For proteins in far-UV CD, the **mean residue weight (MRW)** for the peptide bond is used to normalize the data, calculated as **MRW = M / (N - 1)**, where **M is the molecular mass** of the polypeptide chain (in Da), and **N is the number of amino acid residues**.

Note

Applications of circular dichroism

Circular dichroism (CD) is a highly valuable technique used **to investigate the conformations adopted by proteins and nucleic acids** in solution. Although it doesn't offer the same level of detailed residue-specific information as techniques like NMR and X-ray crystallography, CD analysis boasts several advantages. One of its strengths is the **ability to study molecules of any size**, even **at very low concentrations**. Additionally, CD allows us to examine dynamic systems and kinetics, making it suitable for monitoring structural alterations resulting from changes in environmental conditions such as pH, temperature, and ionic strength.

However, it is important to note that CD provides qualitative analysis rather than atomic-level structural details. The observed spectrum does not exclusively determine a single structure, as multiple possibilities can fit the data.

Nonetheless, CD has a diverse range of applications in the study on biomolecules, particularly proteins and nucleic acids. Some significant applications include:

1. Determining the **secondary and tertiary structure of proteins**.

2. Comparing the **secondary and tertiary structures of wild-type and mutant proteins**.

3. Investigating the **structure of nucleic acids** and changes that occur upon protein binding or melting.

4. Understanding **conformational changes** resulting **from protein-protein interactions, protein-DNA interactions**, and **protein-ligand interactions**.

Analysis of protein's structural and conformational changes

Circular dichroism (CD) is a powerful technique that exploits the differential absorption of left and right circularly polarized light by **chromophores**. This light, depending on its wavelength, is categorized into **far-UV CD (190-250 nm), near-UV CD (250-320 nm)**, and **visible CD** regions. Proteins harbor various chromophores that contribute to CD signals, including the **peptide bond, aromatic amino acid side chains** (such as **tyrosine, tryptophan**, and **phenylalanine**), and **disulfide bonds**. The **far-UV region (240 nm to 190 nm)** reveals crucial information about the protein's overall secondary structure, allowing estimation of the fractions of **alpha-helix, beta-sheet**, and **random coil** conformations.

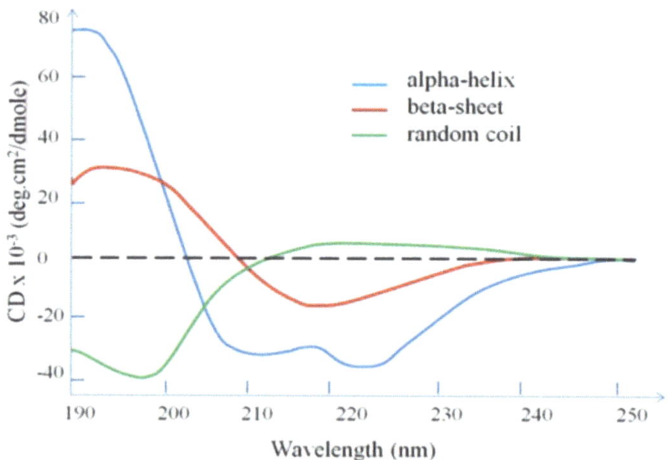

Apart from protein-specific chromophores, **non-protein cofactors** also absorb light across a broad spectral range. Examples include pyridoxal phosphate (~330 nm), flavins (300 nm to 500 nm, depending on oxidation state), heme groups (strong absorption around 410 nm, with other bands between 350 nm to 650 nm, contingent on spin state and coordination of the central Fe ion), and chlorophyll moieties in the visible and near-infrared regions. CD spectroscopy, particularly in the **far-UV region**, offers valuable quantitative insights into a **protein's secondary structure**, facilitating a deeper understanding of its conformational properties.

The **near-UV circular dichroism (CD) spectrum**, measured in the range of 320-260 nm, offers valuable insights into a **protein's tertiary structure**. This spectrum primarily reflects the absorption of light by **aromatic amino acids** and **disulfide bonds**. When a protein possesses a well-defined three-dimensional structure, significant near-UV signals are observed. Conversely, if a protein retains a secondary structure but lacks a distinct three-dimensional arrangement (like a molten-globule structure), the signals will be nearly negligible. Additionally, near-UV CD can be employed **to study the structural details of prosthetic group binding within proteins**.

Materials and Equipment

1. **CD Spectrophotometer**: High-quality CD spectrophotometer equipped with a **Peltier temperature controller** and compatible software for data acquisition and analysis.
2. **Sample Cells**: Quartz cuvettes with a path length of 0.1 cm to 1 cm, suitable for the concentration of your sample.
3. **Sample Preparation**: Your biomolecule of interest (e.g., protein, nucleic acid, peptide) is purified to homogeneity and properly buffer-exchanged to your desired buffer. Use a spectrophotometer to determine the sample concentration and ensure uniformity in subsequent experiments. Prepare reference cuvettes with buffer alone to subtract background noise during data analysis.
4. **Buffer Solutions**: High-quality buffers compatible with your biomolecule. Commonly used buffers include phosphate, Tris-HCl, and HEPES.
5. **Chemicals**: Analytical-grade reagents and solvents are required for sample preparation and buffer preparation.
6. **Disposable Plastic Cuvettes**: For preparing dilutions and testing samples before using precious quartz cuvettes.
7. **Pipettes and Tips**: Adjustable-volume pipettes and disposable tips for accurate sample and buffer dispensing.
8. **Parafilm or Sealant**: To cover sample cuvettes and prevent evaporation during measurements.
9. **Stirrer or Vortex Mixer**: For efficient mixing of samples and buffers.
10. **Deionized Water**: To prepare buffer solutions and dilute samples. A Teflon-coated magnetic stir bar for gentle sample mixing is **optional**.

Protocol

1. **Instrument Preparation**: Turn on the CD spectrophotometer and the Peltier temperature controller. Allow them to warm up for at least 30 minutes before calibration. Check the instrument's wavelength range and ensure it covers the desired spectral range for your experiment.

2. **Sample Preparation**: Ensure your biomolecule is correctly purified and buffer-exchanged. Determine the sample concentration using a spectrophotometer at an appropriate wavelength. Prepare a series of dilutions to find the optimal concentration for your measurements. Fill a disposable plastic cuvette with the sample and test it with the CD spectrophotometer to check the signal quality before using a quartz cuvette.

3. **Baseline Measurement**: Place a quartz cuvette filled with the buffer in the sample compartment. Perform a baseline measurement by scanning the wavelength range of interest (e.g., 190 nm to 260 nm) without any sample.

4. **Sample Measurement**: After obtaining a satisfactory baseline, remove the buffer cuvette, and replace it with your sample-filled quartz cuvette. Ensure the cuvette is clean and free from dust or fingerprints that may interfere with the signal. Record the CD spectrum of your sample in the desired temperature range (if applicable) and wavelength range.

5. **Data Analysis**: Export the raw data obtained from the CD spectrophotometer to compatible software for data analysis. Subtract the baseline spectrum (buffer) from the sample spectrum to obtain the net CD spectrum. Convert the CD signal to mean residue ellipticity (MRE) using the equation: $[\theta] = \theta / (10 \times c \times l)$, where θ is the CD signal, c is the molar concentration, and l is the path length of the cuvette.

6. **Secondary Structure Analysis**: Use CD analysis software to deconvolute the MRE data and determine the percentage of secondary structure elements (α-helix, β-sheet, random coil, etc.). Compare the obtained results with known reference spectra of well-characterized secondary structures.

7. **Conclusion**: Interpret the CD data to conclude the secondary structure and stability of your biomolecule. Discuss the significance of your findings in the context of your research objectives.

Emission spectroscopy

When light is absorbed, atoms or molecules get boosted from their normal state to a higher energy level called the excited state. The excited state can then return to a lower energy state through **two processes: nonradiative transition, where the energy is dissipated as heat**, and **radiative transition, where the energy is released as light**. This light emission following light absorption is known as **photoluminescence**, and the graph of emission intensity versus wavelength is called an **emission spectrum**. Photoluminescence is divided into two categories: **fluorescence**, which involves emission from singlet excited states to singlet ground states, and **phosphorescence**.

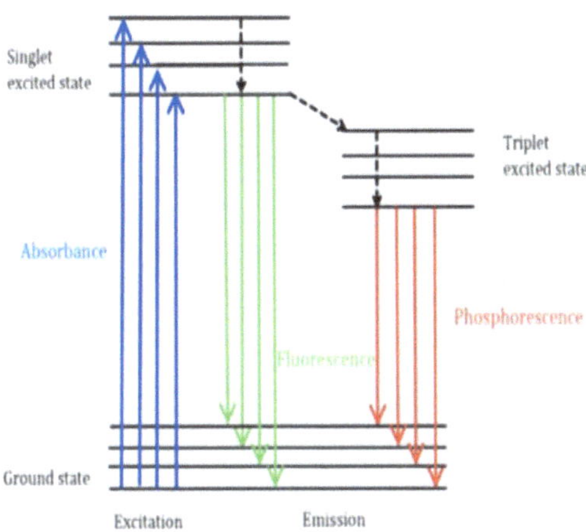

Fluorescence spectroscopy, a type of emission spectroscopy, focuses on **studying the emitted light's wavelengths** when atoms or molecules transition from an excited state to a lower energy state with the same spin multiplicity. Key players in fluorescence are **fluorophores**, which **absorb specific wavelengths and then re-emit the energy at different but equally specific wavelengths**. The emitted energy's amount and wavelength depend on both the fluorophore and its chemical surroundings.

When light interacts with matter, it can be absorbed, scattered, or pass through unaffected. Scattering, a physical process, causes light to change direction from its original path without being absorbed. Unlike absorption, which makes specific wavelengths disappear, both absorption and scattering reduce the transmitted light intensity when passing through a substance. Scattering spectroscopy measures certain physical properties by assessing how much light a substance scatters at specific wavelengths.

Raman spectroscopy

Raman spectroscopy is a powerful non-destructive analytical technique used **to investigate vibrational, rotational, and other low-frequency modes in a sample**. This technique **relies on the inelastic scattering of monochromatic light,** typically from a **laser source**, to probe the vibrational energy levels of the sample molecules. Raman spectroscopy provides valuable information about the **chemical composition, molecular structure**, and **crystalline nature** of a wide range of materials, including solids, liquids, and gases.

Fundamental concept

Raman spectroscopy is a powerful technique used **to study the interaction between light and matter**. Unlike absorption or emission processes, Raman spectroscopy **involves the scattering of radiation by a sample**. When light interacts with a molecule, it can lead to **absorption or scattering**. Absorption occurs when a photon's energy matches the difference between two energy levels in the molecule, causing a transition from a lower to a higher energy state. However, in Raman scattering, the incident light is inelastically scattered at frequencies different from the incident light, involving a net energy transfer between the incident photons and the material. This unique feature allows researchers to gain valuable insights into molecular vibrations and structural information without significant absorption or emission effects, making Raman spectroscopy an essential tool in various scientific fields.

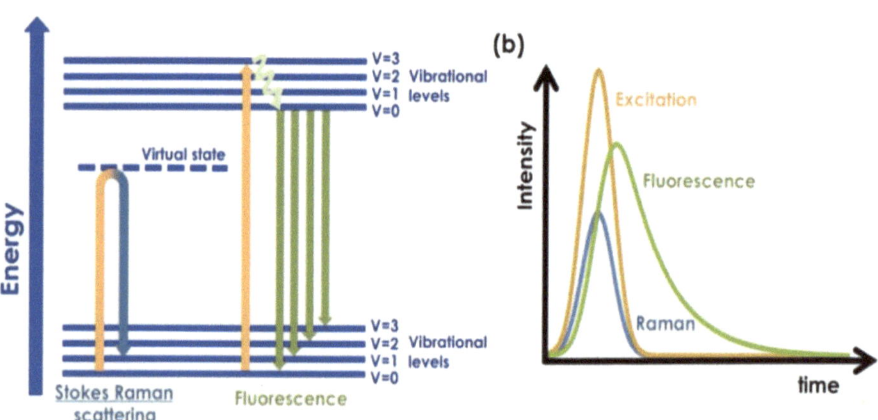

Raman spectroscopy is a powerful analytical technique that involves the **scattering of light by molecules**. When a photon interacts with a molecule, its energy changes, leading to either longer wavelengths (**Stokes Raman scattering**) or shorter wavelengths (**anti-Stokes Raman scattering**) scattered photons. The energy of scattered radiation is lower for Stokes lines and higher for anti-Stokes lines compared to the incident radiation. This energy difference, known as the **Raman shift**, corresponds to the variance in vibrational energy levels within the molecule, allowing us to identify different vibrational modes through characteristic **Raman shifts** (or 'bands') in the scattered light spectrum. Both Raman and infrared (IR) spectroscopy rely on molecular vibrations associated with chemical bonds to provide valuable information about a sample's molecular structure, composition, and intermolecular interactions. While **IR spectroscopy** measures the net **absorption of incident radiation** by the sample in the infrared region, **Raman spectroscopy** examines **vibrational and rotational motion** resulting from inelastic scattering events influenced by nuclear vibrations, which affect the molecule's polarizability during its vibrational and rotational movements.

Raman spectroscopy, unlike infrared absorption, does not require matching the incident radiation to the energy difference between the molecule's ground and excited states. Instead, it involves the interaction of light with the molecule, leading to a temporary state known as a 'virtual state.' In this process, the cloud of electrons surrounding the molecule is distorted (polarized), and the photon is rapidly re-emitted. By studying the intensity profiles of scattered light as a function of frequency, Raman spectroscopy provides valuable insights into the frequency of vibration of the molecular bonds. This unique ability to explore molecular vibrations and rotations makes Raman spectroscopy a valuable tool in scientific research and various fields of study.

Virtual learning aid

Note

Materials and Equipment

1. **Raman Spectrometer**: High-quality, commercially available Raman spectrometer equipped with a laser light source, detector, and data acquisition system.
2. **Laser Source**: Monochromatic laser with appropriate wavelength and power (commonly used wavelengths include 532 nm, 785 nm, and 1064 nm).
3. **Sample Holder**: A suitable sample holder or cuvette that allows for proper positioning and stability of the sample during analysis.
4. **Sample Substrates**: Clean glass slides or suitable substrates for supporting liquid or solid samples.
5. **Optics**: Collection optics (lenses or mirrors) to focus the laser on the sample and collect the Raman-scattered light.
6. **Calibration Standard**: Raman-active reference material for instrument calibration (e.g., silicon or polystyrene).
7. **Computer and Software**: A computer with appropriate software for data acquisition and spectral analysis.

Protocol

1. **Sample Preparation**:
A. For Solids: Grind the sample to a fine powder using a mortar and pestle to ensure uniformity and avoid scattering variations. Place a small amount of the powder on a clean glass slide or substrate.
B. For Liquids: Use a dropper to place a small drop of the liquid sample onto a clean glass slide or into a suitable cuvette.

2. **Instrument Setup**: Turn on the Raman spectrometer and allow it to warm up for the specified time (consult the instrument manual). Align the laser source with the sample holder or cuvette to ensure proper focusing. Use the calibration standard to calibrate the instrument following the manufacturer's guidelines.

3. **Sample Analysis**: Position the sample holder with the prepared sample in the spectrometer's sample chamber. Close the chamber and ensure that the sample is appropriately aligned with the laser beam. Set the laser power to the recommended level to avoid sample damage. Start the data acquisition software and select the appropriate acquisition parameters (e.g., integration time, spectral range, and resolution). Acquire the Raman spectrum by clicking the "Start" button in the software.

4. **Data Analysis**: Save the acquired Raman spectra in a suitable format for further analysis. Use the software to perform baseline correction and data processing, if required. Compare the obtained spectra with reference databases or known spectra to identify the sample's chemical composition and molecular structure. Interpret Raman peaks in terms of bond vibrations and molecular modes using Raman databases or literature.

Chapter 7

Centrifugation

Introduction

Centrifugation is a powerful technique used **to separate or concentrate materials suspended in a liquid by applying centrifugal force**. This force is generated by spinning the liquid at high speeds in a device called a **centrifuge**. The key to separation lies in the differences in the sedimentation rate of particles under this centrifugal field. Various factors, such as **size, density**, and **shape**, influence how quickly particles settle. When particles are suspended in a solution, they are pulled downward by Earth's gravity. However, for very small particles, the weak gravitational force is usually insufficient to cause significant sedimentation due to random thermal motion. Centrifugation overcomes this limitation by creating forces much stronger than gravity, often up to 10,00,000 times stronger. By doing so, it enhances the sedimentation of these smaller particles.

To explain the principle, let's consider a solution being spun in a centrifuge tube. The centrifugal force acting on a particle with **mass 'm'** is given by $\mathbf{ma^2r}$, where **'a' is the angular velocity in radians per second**, and **'r' is the distance from the center of rotation to the particle**. This force causes the particles to move through the liquid medium.

Virtual learning aid

However, it's essential to consider the particle's buoyancy due to the solvent displacement. The solvent molecules displaced by the particle create a buoyancy that reduces the effective force on the particle by w^2r times the mass of the displaced solvent. Centrifugation serves both preparative and analytical purposes. **Preparative centrifugation**, separates one type of material from others, while analytical centrifugation is employed to measure physical properties like molecular weight, density, and shape.

Relative Centrifugal Field

Centrifugation is a process that separates particles suspended in a fluid by using centrifugal forces, which are much stronger than Earth's gravity. When a vessel containing the fluid and particles is spun around an axis of rotation, the particles experience a force called **relative centrifugal force (RCF)** or "g force." RCF is expressed as a multiple of Earth's gravitational force. For instance, RCF 500 g means the centrifugal force is 500 times stronger than Earth's gravity.

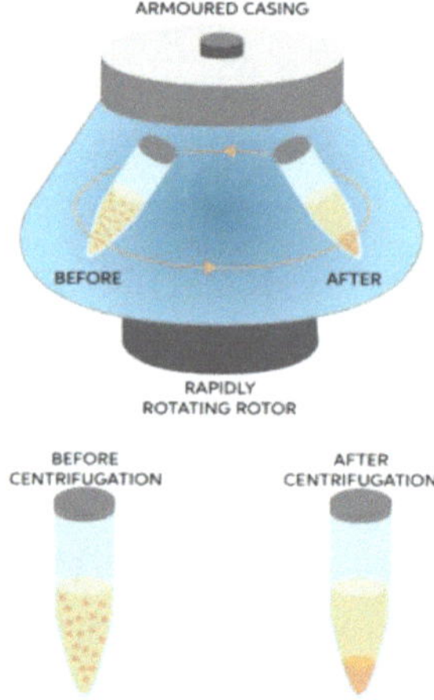

The RCF generated in a centrifuge depends on the **speed of the rotor** (measured in revolutions per minute or rpm) and the **radius of rotation** (the distance from the axis of rotation). The relationship can be calculated using the equation:
RCF = $(r \times \omega^2)$ / g
Where,
1. **r** is the **radius in centimeters**
2. ω is the **angular velocity in radians per second**
3. **g** is the **gravitational acceleration in centimeters per second squared.**
The **angular velocity (ω)** can be determined using the formula: $\omega = 2\pi \times \textbf{rpm}$ **/ 60**. For example, at 2000 rpm, ω would be approximately 209.44 radians per

second. Combining this with the radius and gravitational acceleration, we can calculate the RCF for a given centrifuge.

RCF serves as a measure of the strength of the centrifugal field created by different-sized rotors and operating speeds. A larger centrifuge with a greater radius of rotation will spin samples at a higher RCF compared to a smaller centrifuge with a shorter radius, even at the same rpm.

Virtual learning aid

Fixed-angle or swing-bucket rotors

In laboratory settings, **two** commonly used rotor types are the **fixed-angle** and **swing-bucket rotors**. These rotors differ in their geometry and, as a result, exhibit distinct sedimentation properties.

Differential centrifugation

Differential centrifugation is a basic centrifugation technique that separates particles in a suspension **based on their different sizes or densities**. By applying centrifugal force, the larger and denser particles settle faster, followed by those with intermediate sizes or densities, and finally, the smallest or least

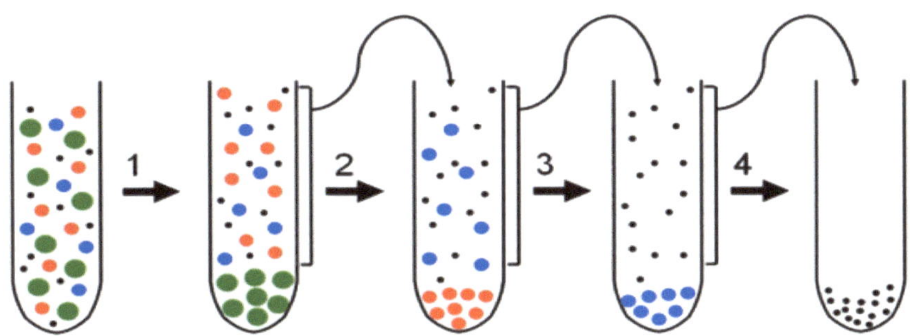

dense particles. This method is commonly used **to separate sub-cellular components**, and eventually, all particles settle at the bottom of the tube.

Separation of sub-cellular components by differential centrifugation

During the process of sub-cellular component separation, broken cells are centrifuged at progressively higher forces and durations. Initially, the largest and heaviest sub-cellular component is completely sedimented at low centrifugal force. The supernatant is then decanted and subjected to higher centrifugal force to sediment the next heavier entity. This step-by-step approach continues, leading to the sedimentation of lighter components based on their differential sizes and densities. Ultimately, the procedure yields separate fractions, including **nuclei** and unbroken cells, **mitochondria**, **chloroplasts**, **lysosomes**, **peroxisomes, plasma membrane**, and **endoplasmic reticulum fragments**, as well as the soluble portion of **cytoplasm** (cytosol).

Virtual learning aid

Materials and Equipment

1. Microcentrifuge tubes (1.5 mL and 15 mL)
2. Refrigerated centrifuge with appropriate rotor and adapters
3. Homogenizer (Dounce homogenizer or mechanical homogenizer)
4. Ice bucket or container with ice
5. Centrifuge tube racks
6. Pipettes and pipette tips
7. Phosphate-buffered saline (PBS) or appropriate buffer

8. Protease and phosphatase inhibitors (optional but recommended)
9. Sterile water
10. Cell scraper (optional)
11. Benchtop vortex mixer
12. Balance

Protocol

(Note: Perform all steps **at 4°C** unless otherwise specified.)

1. **Sample Preparation**: Harvest the cells or tissues of interest and wash them gently with cold PBS to remove any debris and residual media. Centrifuge the suspension at 300 x g for 5 minutes to pellet the cells. Discard the supernatant carefully. Resuspend the cell pellet in an appropriate volume of cold PBS. The volume will depend on the cell type and starting material. Optionally, add protease and phosphatase inhibitors to the resuspended cells to minimize protein degradation.

2. **Cell Disruption and Homogenization**: Transfer the cell suspension to a pre-chilled homogenization vessel (Dounce homogenizer or mechanical homogenizer). Homogenize the cells gently to disrupt the cell membrane and release cellular components. Avoid excessive force to prevent organelle damage. Verify the effectiveness of cell disruption by viewing a small aliquot under a microscope.

3. **Centrifugation Step 1 (Low-Speed Centrifugation)**: Transfer the homogenate to a 15 mL centrifuge tube and balance it with sterile water in case necessary. Centrifuge the tube at 800 x g for 10 minutes to pellet unbroken cells, nuclei, and large debris. This pellet is called the "P1" fraction. Carefully decant the supernatant (S1) into a fresh 15 mL centrifuge tube and keep it on ice.

4. **Centrifugation Step 2 (Medium-Speed Centrifugation)**: Centrifuge the S1 fraction at 10,000 x g for 20 minutes. This step separates the crude mitochondria and other organelles from the cytosolic fraction. Collect the supernatant (S2) carefully, as it contains the cytosolic fraction. Resuspend the pellet (P2) containing the crude organelles in a suitable volume of cold PBS or buffer.

5. **Centrifugation Step 3 (High-Speed Centrifugation)**: Take the S2 fraction and centrifuge it at 100,000 x g for 60 minutes. This separates the lighter microsomes from the heavier organelles like the endoplasmic reticulum. Collect the supernatant (S3), which contains the microsomal fraction. The resulting pellet (P3) contains heavier organelles, such as the endoplasmic reticulum. Resuspend this pellet in a cold buffer.

6. **Optional Ultracentrifugation Step (For Further Separation)**: If required, subject the S3 fraction to an additional ultracentrifugation step at a higher speed (e.g., 150,000 x g) to isolate specific membrane fractions. Collect the supernatant, which will contain the lighter membrane fractions. The resulting pellet may contain specialized organelles or membrane complexes of interest.

Density gradient centrifugation

Differential centrifugation involves separating particles initially uniformly distributed in a solution by subjecting them to centrifugal force. **Density gradient centrifugation**, a variation of this method, **employs a medium with increasing density from top to bottom**. This gradient can form either by preforming it (**rate zonal centrifugation**) or by self-generating during centrifugation (**isopycnic centrifugation**).

Virtual learning aid

Rate zonal centrifugation

Rate zonal centrifugation, also known as **velocity centrifugation**, is a powerful technique used **to separate particles based on their differing sedimentation rates**. This method effectively separates particles according to their **size or density**, with even subtle variations in shape influencing the sedimentation rate.

To prepare density gradients in this process, materials like **sucrose**, **glycerol**, and **ficoll** are commonly used. A 5-20% sucrose solution forms the density gradient, carefully selected to ensure that particles' density exceeds that of the medium at all points during the separation. The density gradient plays a crucial role by stabilizing particle bands and enhancing resolution through the inhibition of convection currents.

During the Rate zonal centrifugation, the samples are centrifuged for a precise duration, separating the molecules of interest into discrete zones. If the centrifugation time exceeds the necessary duration, all components of the sample will collect at the tube's bottom in a pellet due to the particles' higher density than the gradient. Apart from preparing and purifying macromolecules and cellular components, this centrifugation technique also enables the determination of sedimentation coefficients and molecular weights of biological macromolecules. By subjecting a purified molecule, such as a protein, to a centrifugal field, the molecule will sediment toward the bottom at a constant velocity. At this stage, the **molecular weight (M)** can be calculated using the equation:

$M = f \times v/\omega 2r$, where **'f' represents the frictional coefficient** of the solvent system (determined from other measurements), and **'v/ω2r' corresponds to the sedimentation coefficient.**

Virtual learning aid

Materials

1. Centrifuge
2. Centrifuge tubes
3. Sample
4. Gradient medium (sucrose or cesium chloride)
5. Buffer solution
6. Pestle or homogenizer
7. Microcentrifuge tubes
8. Pipettes and pipette tips
9. UV spectrophotometer
10. Centrifuge safety gear (lab coat, gloves, goggles)

Protocol

1. **Preparation of Gradient Medium**: Prepare the **buffer solution** according to the specific requirements of your experiment. The buffer should maintain the stability and integrity of the biomolecules during centrifugation. Calculate the

required concentration of the **gradient medium** based on the properties of the molecules you want to separate. Prepare the gradient medium by dissolving the appropriate amount of sucrose or cesium chloride in the buffer solution. Ensure the solution is free from air bubbles.

2. **Sample Preparation**: Homogenize or disrupt the biological sample to release the biomolecules of interest. Use a pestle or homogenizer, and work on ice or in a cold room to prevent sample degradation. Centrifuge the homogenized sample at low speed to remove cellular debris or any large particles. Collect the supernatant, which contains the biomolecules to be separated.

3. **Preparing the Gradient Tube**: Choose a clear and graduated centrifuge tube appropriate for the volume of your sample. Carefully layer the gradient medium into the tube using a long, fine-tipped pipette to avoid mixing with the sample. Gently overlay the sample onto the gradient medium, being cautious not to disturb the gradient layers.

4. **Centrifugation**: Load the filled centrifuge tubes into the rotor following the manufacturer's guidelines to maintain balance. Set the appropriate centrifugation speed (in RPM) and time (in minutes) based on the properties of the molecules you want to separate. Start the centrifugation process, ensuring that the tubes are well-secured within the rotor to prevent any leakage or breakage.

5. **Fraction Collection**: After centrifugation, carefully remove the centrifuge tubes from the rotor. Using a microcentrifuge tube, collect fractions of interest from the top of the tube, carefully avoiding cross-contamination between fractions.

6. **Analysis**: Analyze the collected fractions using suitable methods like UV spectrophotometry or electrophoresis to identify the separated biomolecules. Plot the results to visualize the distribution of molecules across the gradient.

Isopycnic centrifugation

Isopycnic centrifugation, also known as **buoyant density centrifugation** or **equilibrium density-gradient centrifugation**, offers a unique approach to separate particles **based solely on their buoyant density, disregarding their shape and size**. Unlike rate zonal centrifugation, isopycnic centrifugation's efficiency is unaffected by the centrifugation time. This technique proves particularly useful for segregating particles with similar sizes but varying densities. During the process, particles progressively settle in a density gradient until their buoyant density matches that of the gradient. Typically, a steep density gradient containing **sucrose (20-70%)** or **CsCl** is employed for this purpose.

Materials and Equipment

1. **Gradient Medium**: Prepare a density gradient medium using **sucrose, cesium chloride (CsCl),** or another suitable material. The concentration and volume required will depend on your specific experiment.

2. **Sample**: The sample you want to separate should be collected and prepared beforehand. Ensure that it is properly homogenized and free from any particulate matter.

3. **Centrifuge Tubes**: High-quality, clean, and dry centrifuge tubes suitable for the volume of the sample and the rotor capacity.

4. **Ultracentrifuge**: An ultracentrifuge capable of achieving high speeds and generating sufficient centrifugal force is essential.

5. **Rotor**: Select an appropriate rotor that can accommodate your centrifuge tubes securely.

6. **Graduated Pipettes and Tips**: To measure and transfer precise volumes sample and gradient medium.

7. **Centrifuge Tube Adapters**: If needed, use adapters to ensure proper fitting of tubes in the rotor.

8. **Balance**: A high-precision balance to weigh the sample and gradient medium accurately.

9. **Centrifuge Safety Accessories**: Wear appropriate personal protective equipment (PPE) such as lab coats, gloves, and safety goggles.

Protocol

1. **Preparation of Gradient Medium**: Calculate the density and volume of the gradient medium required for your experiment based on the sample and separation objectives. Dissolve the appropriate amount of sucrose or cesium chloride in an appropriate buffer to form the gradient medium. Mix thoroughly until the solution is clear and homogeneous.

2. **Sample Preparation**: Collect the sample and homogenize it carefully to ensure an even distribution of components. Filter the sample through a fine mesh or centrifuge it at a low speed to remove large particulate matter and debris. Measure the volume of the sample accurately.

3. **Loading the Gradient Tube**: Take a clean and dry centrifuge tube. Gently layer the prepared gradient medium at the bottom of the tube using a graduated pipette. Avoid introducing air bubbles into the gradient.

4. **Layering the Sample**: Carefully layer the prepared sample on top of the gradient medium using a separate graduated pipette. Slowly pipette the sample onto the gradient medium to form distinct layers without mixing the two.

5. **Centrifugation**: Place the loaded centrifuge tubes securely in the rotor, making sure they are balanced. Centrifuge the tubes at the specified speed and duration required for your separation.

6. **Harvesting the Fractions**: After centrifugation, carefully remove the tubes from the rotor. Open the tubes and collect fractions from different layers of the gradient using a pipette. Label the fractions properly for identification.

7. **Analysis of Fractions**: Analyze each collected fraction using suitable analytical methods such as **spectrophotometry**, **chromatography**, or **electrophoresis**, depending on the nature of your sample.

Note

Rate zonal centrifugation separates particles based on their sedimentation coefficients, while isopycnic centrifugation separates particles based on their buoyant density.

When centrifuged in a CsCl solution, macromolecules form bands at specific points in the gradient based on their buoyant density. **DNA**, with a buoyant density of approximately **1.70 g/cm3**, migrates to the point in the gradient where the CsCl density matches. In contrast, **proteins**, having **lower buoyant densities**, remain **at the top**, while **RNA**, with **higher buoyant density**, forms a **pellet at the bottom**. Hence, isopycnic centrifugation effectively separates DNA, RNA, and proteins.

DNA separation by isopycnic centrifugation

Isopycnic centrifugation employing CsCl is a valuable technique for fractionating and quantitatively separating DNA molecules based on their buoyant densities. CsCl, with a density of **1.6 to 1.8 g/mL**, is preferred due to its similarity to DNA density. Most linear double-stranded DNA in CsCl solution has a buoyant density of **1.70 g/cm3**, while **single-stranded DNA** exhibits a slightly higher density, about **0.015-0.020 g/cm3 greater than double-stranded DNA** of the same base composition. Additionally, the buoyant density of **single-stranded RNA** in CsCl solution **exceeds 1.8 g/cm3**.

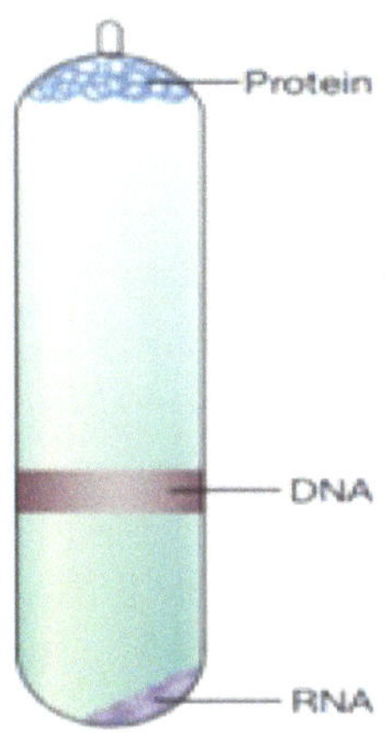

Separation based on differences in GC content

DNA molecules with varying proportions of AT base pairs to GC base pairs can be separated using CsCl density gradient centrifugation. Since AT base pairs have a lower molecular weight than GC base pairs, DNA molecules with **higher AT** content will have a **lower buoyant density** if they are of equal length. This creates a linear relationship between the buoyant densities of DNA and their GC content.

Buoyant density (p) = 1.660 +0.098 (G+C) g/cm3.

In scientific studies, researchers employ CsCl density gradient centrifugation to distinguish **highly repetitive satellite DNA** from the remaining DNA. Satellite DNA, which consists of numerous short tandem repeats, may exhibit a distinct base composition and GC content compared to the whole genome. Consequently, satellite DNA demonstrates a different buoyant density from the rest of the DNA due to this variation in base composition. Through CsCl equilibrium density gradient centrifugation, DNA is fractionated based on its density, resulting in separate bands. When the GC content varies by 5% or more, distinct bands emerge. For instance, in a study involving mouse nuclear DNA fragments, two DNA bands appeared - one containing 92% of the DNA with a density of 1.701 gm/cm3 (GC content ~42%) and a thinner satellite band containing 8% of the DNA with a density of 1.690 gm/cm3 (GC content ~30%). However, in cases where the average satellite DNA base composition closely resembles that of the entire genome, physical separation using a density gradient becomes unfeasible.

Note

Ethidium Bromide (EtBr) is a flat molecule that inserts itself between the base pairs of the DNA double helix. This insertion causes the double helix to unwind, leading to a decrease in the **twist (Tw)** and an increase in the **writhe (Wr)** on the DNA molecule. Remarkably, the insertion of just one EtBr molecule results in the untwisting of the helix by 26°. However, it takes the addition of 14 EtBr molecules to induce the unwinding of an entire turn of the DNA helix, causing the Tw to decrease by one.

Separation based on conformation

In the laboratory, scientists use a technique called CsCl equilibrium density gradient centrifugation **to separate linear DNA (relaxed) molecules from circular supercoiled DNA**. This process involves the presence of a DNA intercalating dye called EtBr. When EtBr intercalates with DNA, it unwinds the DNA helix, reducing the buoyant density of linear DNA by approximately 0.125 g/cm3. However, circular supercoiled DNA, which lacks free ends, has limited flexibility to unwind and can only bind a limited amount of EtBr. Consequently, the buoyant density of circular supercoiled DNA decreases by around 0.085 g/cm3, significantly less than linear DNA. This difference in buoyant density allows circular supercoiled DNA to form a distinct band in an EtBr-CsCl gradient, enabling its separation from linear DNA.

Chapter 8

Microscopy

Introduction

Microscopy is a method that enables us **to observe extremely small objects that are not visible to the naked eye**. It utilizes instruments known as **microscopes**, which come in **two** main types: **light microscopes** and **electron microscopes**. These microscopes differ in the way they illuminate the specimen, allowing us to explore the microcosmic world with greater detail and clarity.

Light microscope

The **light microscope**, also known as the **optical or transmission light microscope**, uses visible light to illuminate specimens. It consists of **lenses**, such as a single magnifying glass in a simple form or multiple lenses in a compound microscope. The **compound microscope** includes the **condenser lens, objective lens**, and **eyepiece lens**. The **condenser lens focuses light** from the source onto the specimen, while the **objective lens**, with different magnification options (e.g., 4x, 10x, 40x), produces the **initial magnified image**. The **eyepiece lens** further magnifies this image, usually by 10x, resulting in a **final magnified** and **virtual image** viewed by the eye.

In the **compound microscope**, the objective lens forms a **real** and **inverted magnified image** of the object, creating a real intermediate image in the focal plane of the eyepiece. This intermediate image serves as the object for the eyepiece, which magnifies it to produce an erect and further magnified virtual image. Though inverted compared to the object, the virtual image observed by the eye is clear and enlarged. The eye's lens forms a real image on the retina since the virtual image from the eyepiece is well outside the focal length of the eye. This combination of lenses allows scientists and researchers to observe minute details of specimens and study various biological structures and materials.

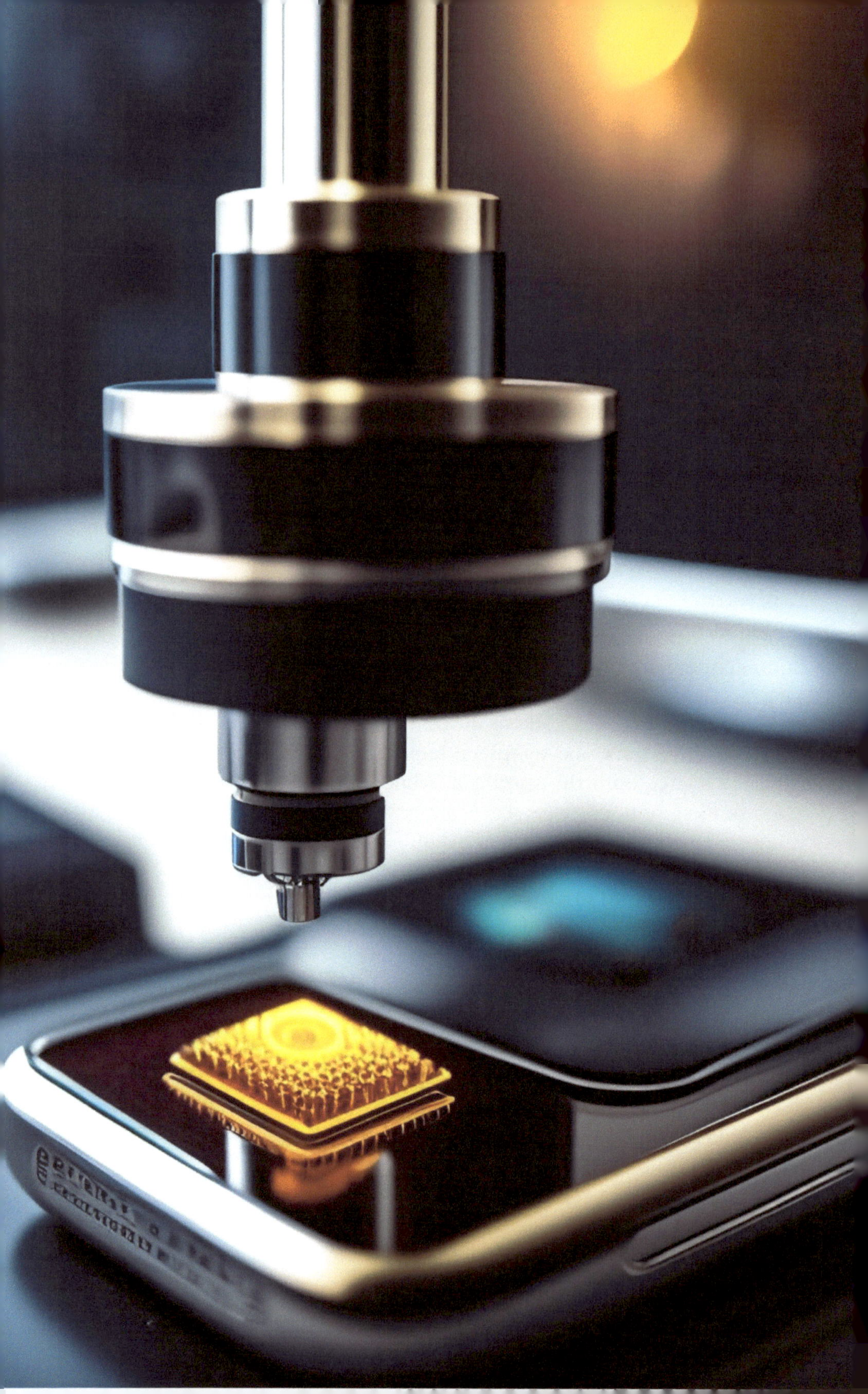

Magnification

The **magnification** of a compound microscope is the result of multiplying the magnification of the objective lens with that of the eyepiece. The **objective lens** contributes to the **linear magnification** of the microscope, which is the **ratio of the image size to the object size**, measured in linear dimensions.

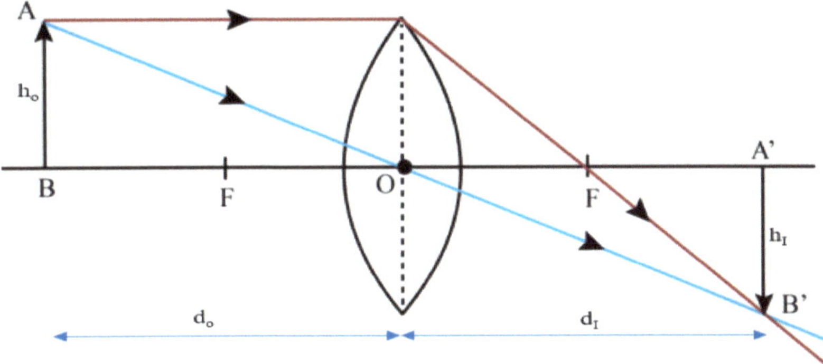

Linear magnification is a measure of how much larger or smaller an image appears compared to the actual object. It is defined as the ratio of the image size to the object size and can also be expressed as the **ratio of the distance of the image from the lens to the distance of the object from the lens**.

Mathematically, it is represented as $m = h_I/h_o = d_I/d_o$, where **h_I is the height of the image** formed, **h_o is the height of the object** being observed, **d_I is the distance of the image** formed from the lens, and **d_o is the distance of the object from the lens**.

Note

Magnification and resolution are crucial aspects of microscopy. Magnification refers to how much larger an image appears compared to the actual size of the specimen. On the other hand, resolution denotes the smallest distance at which two discrete objects can be seen as distinct entities. Magnification depends on the number of lenses used, while resolution is influenced by a lens's ability to gather light. For compound light microscopes, the maximum magnification typically reaches around 1500x, with a resolution limit of approximately 0.2 μm.

When discussing microscopes, we often encounter the term "**angular magnification**" or "**magnifying power.**" This refers to the magnification produced by the eyepiece, and it represents the **ratio between the angle subtended by the image on the eye when using the microscope and the angle subtended by the object when viewed without the microscope** (i.e., with the naked eye).

Angular Magnification = Angle subtended by image/Angle subtended by the object

The **overall magnification** of a lens can be calculated by using both the **linear magnification (Ml)** and the **angular magnification (Ma)** values. The formula for overall magnification (M) is given by:

M = Ml × Ma

Where,

1. **M is the overall magnification** of the lens,

2. **Ml is the linear magnification**, which is the ratio of the size of the image to the size of the object,

3. **Ma is the angular magnification**, which is the ratio of the angle subtended by the image when viewed through the lens to the angle subtended by the object when viewed without the lens.

Virtual learning aid

Amplitude contrast and phase contrast

When using a light microscope to observe specimens, both living and dead samples are examined. The **visibility of the magnified specimen** is determined **by contrast** and **resolution**. **Contrast** refers to the **differences in light intensity within an image** and can be of **two** types: **amplitude contrast**, which relies on **differences in the amplitude of transmitted light**, and **phase contrast**, which depends on the **phase difference of transmitted light**. Since light intensity is related to amplitude and proportional to the square of its amplitude, specimens seen under transmission light microscopy are classified into amplitude objects (e.g., stained specimens) and phase objects (e.g., transparent colorless objects like living cells). Amplitude objects create direct amplitude differences in the image due to differential absorbance of light, leading to color or brightness contrast. On the other hand, phase objects do not absorb light but diffract it, causing a phase shift in the passing rays. Some light microscopes transform these phase differences into amplitude differences in the image.

In the case of colorless transparent specimens, such as most biological materials, achieving contrast poses unique challenges. Staining the specimen can be one approach to reduce the amplitude of specific light waves passing through stained areas, resulting in color contrast. However, this typically requires the fixation and staining of specimens, which can be detrimental as it kills the living specimens. For colorless living specimens, contrast can be achieved through different means. Living biological specimens are almost uniformly transparent, leading to very low-intensity variation due to differential absorption (amplitude contrast). Nevertheless, light passing through the specimen undergoes a phase shift caused by scattering and diffraction, thereby producing phase contrast, which aids in observing these otherwise transparent living samples.

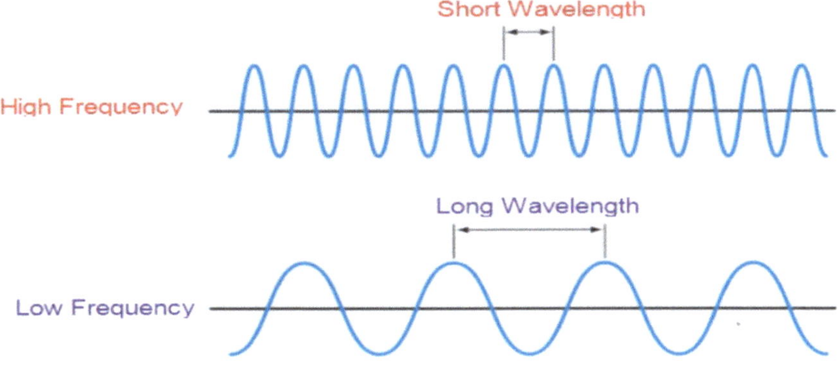

Resolving power

Resolving power refers to the **ability of a magnifying instrument**, such as a microscope, **to distinguish two closely spaced objects**. It is **inversely related to the limit of resolution**, which is the smallest distance between two points that can be seen as separate entities. Essentially, **higher resolving power means a smaller limit of resolution**.

The **limit of resolution** in a light microscope depends on **three factors**: the **wavelength** of the light used to illuminate the specimen, the **angular aperture**, and the **refractive index** of the medium between the cover slip and the objective lens. These factors are quantitatively related through the **Abbe equation**:

Limit of resolution = 0.61λ / (n × sin α)

= 0.61λ / Numerical aperture (NA)

Here,

1. λ represents the **wavelength of the light**

2. **n** is the **refractive index** of the medium between the cover slip and the objective lens

3. α is **half of the angular aperture**.

The **numerical aperture (NA)** of the objective lens measures its **ability to gather light from the specimen**. A **higher NA indicates better light collection**.

The **unaided human** eye has a **limit of resolution** of about **100 µm**. To improve resolution in microscopy, we need to minimize the numerator (λ) in the Abbe equation (using light with the shortest wavelength) and maximize the denominator (numerical aperture). This can be achieved by:

1. **Increasing** the one-half angular aperture (α) of the objective lens.
2. **Increasing** the refractive index (**n**) of the medium between the objective lens and the specimen.
3. **Shortening** the wavelength (λ) of the illuminating light.

By optimizing these factors, we can enhance the resolving power of a microscope and discern finer details in the observed specimen.

Angular aperture

The **angular aperture** refers to the **angle of the cone of light collected by the objective lens** in a microscope. It varies depending on the size and shape of the illumination cone entering the lens. A larger light cone leads to a higher angular aperture. This can be achieved by reducing the distance between the objective lens and the specimen or by increasing the lens diameter. The **maximum possible angular aperture** for a standard microscope objective lens is theoretically **180°**, resulting in a one-half angular aperture value of 90°. However, in practice, the **best objective lenses** typically have a working one-half angular aperture of around **70°**. Therefore, the **maximum value for sin α**, a measure of numerical aperture, is approximately **0.94**. Since the refractive index of air is about 1.0, no lens working in air can have a numerical aperture greater than 1.0.

Refractive index

In microscopy, the **refractive index** of air limits the maximum numerical aperture of lenses to around 1.0. However, some microscope lenses use **immersion oil** with a higher refractive index (about **1.5**) to enhance light reception through the specimen. This enables **improved resolution**, with limits of around **300 nm in air** and **200 nm with immersion oil** when using visible light.

Wavelength of the illuminating light

For optimal resolution, the numerator's minimum value requires a smaller wavelength, such as illuminating the specimen with **blue light at 450 nm**.

Oil immersion lens

In most microscopes, the space between the coverslip protecting the sample and the front lens of the objective is filled with air. This type of objective is called a **dry objective** since it does not require any additional medium. However, air has a refractive index close to that of a vacuum and significantly lower than most liquids like water. To enhance the resolving power of the microscope, **immersion oil** is used. An oil immersion objective is specially designed for this purpose. In this system, the air between the coverslip and the objective lens is replaced with a transparent oil of high refractive index, similar to that of glass. By eliminating the air-glass interface, which causes light refraction and scattering, more light can be directed through the objective, resulting in a clearer image.

Immersion oils, such as **paraffin oil** or **cedarwood oil**, are placed at the interfaces between the objective lens and the coverslip, as well as between the condenser lens and the underside of the specimen slide. This replacement of air with oil greatly reduces refraction and improves resolution in light microscopy. The immersion oil's refractive index matches that of glass, allowing more light to be efficiently directed through the high-power objective lens. Consequently, the use of immersion oil leads to better image clarity and sharper details in the observed specimens.

Types of light microscope

1. **Bright-field microscopy**: **Bright-field microscopy** is the original and most widely used form of microscopy. When light passes through a specimen, it can undergo changes in **amplitude, phase**, and **polarization**. In this technique, **color contrast is generated** by the differential absorption of light by stained specimens or those with natural pigmentation. If specimens are thick enough to absorb a significant amount of light, even if they are colorless, they can still create brightness contrast. The absorption of light reduces the amplitude in transmitted light, resulting in amplitude contrast.

Virtual learning aid

Scan It!

2. Dark-field microscopy: A **dark-field micros**cope is designed **to enhance contrast in unstained, transparent specimens**. By illuminating objects at a low angle from the side, the background appears dark, allowing the specimens to stand out clearly. This is achieved by using a **central circular disk** stop to block direct condenser rays from entering the objective lens. Only the light scattered by the specimen is collected, creating a final image with improved contrast.

Virtual learning aid

3. **Phase-contrast microscopy: Phase-contrast microscopy** is a valuable technique for studying **transparent** and **colorless specimens** without the need for cell killing, fixing, or staining. This method utilizes the fact that when light passes through a living cell, **its phase changes due to variations in cell thickness and density**. For instance, light passing through denser parts of the cell, like the nucleus, experiences a phase shift compared to light that passes through less dense regions in the cytoplasm. By considering two beams of light from the same source, one passing through a transparent, denser material and the other through air, we can observe a phase difference between the two beams. This phase difference, caused by changes in the speed of light passing through different materials, allows us to transform variations in the relative phase of light waves into amplitude differences in the final image, enabling us to visualize the specimen's internal structures without the need for staining or other invasive procedures.

In phase contrast microscopy, when transparent, unstained specimens are used as phase objects, they do not absorb light but rather exhibit diffraction or scattering of light. This diffraction leads to a phase difference between the non-diffracted (direct) and diffracted beams of light. Additional phase differences are introduced during the imaging process, further altering the relative phases of the light waves. Consequently, interference between these sets of beams results in overall differences in amplitude, which are perceived as variations in brightness and contrast in the final image. This powerful technique allows researchers to visualize live cells and other transparent specimens in great detail, providing valuable insights into their structures and dynamics without compromising their viability or integrity.

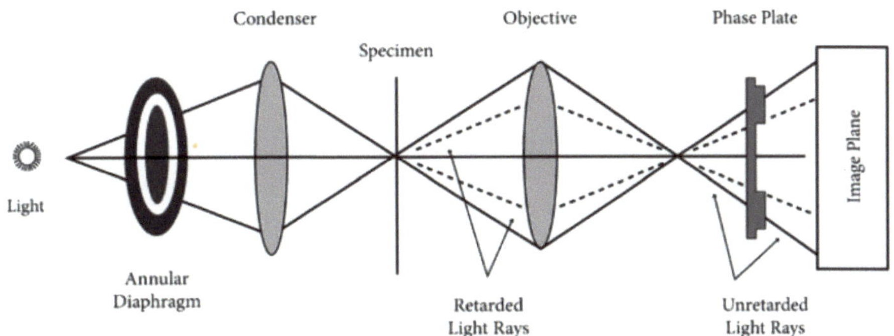

Components of phase contrast microscope

Fluorescence microscope

A fluorescence microscope utilizes specimens that act as light sources themselves. These specimens can be either fluorescent materials or stained with fluorescent dyes. Fluorescent chemicals absorb light at one wavelength and emit light at a longer wavelength. Most fluorescent dyes emit visible light, while some emit infrared light. These dyes have distinct excitation and emission spectra based on their atomic structure and electron resonance properties. Commonly used dyes include **fluorescein, which emits intense green fluorescence** when **excited with blue light**, and **rhodamine, which emits deep red fluorescence** when **excited with green-yellow light**.

A **fluorescence microscope** consists of **three** main components: **an excitation filter**, **a dichroic mirror**, and **an emission or barrier filter**. The excitation filter allows specific wavelengths of light to excite a particular fluorophore in the sample. This filtered excitatory light is then reflected toward the sample by the dichroic mirror, which sits at a **45° angle**. The dichroic mirror enables only the emitted fluorescent light of a longer wavelength from the sample to reach the eyepiece. To block extraneous excitatory light or background fluorescence, an emission or barrier filter is often used. This filter ensures that only the fluorescent light emitted by the sample is allowed to form the image observed through the microscope.

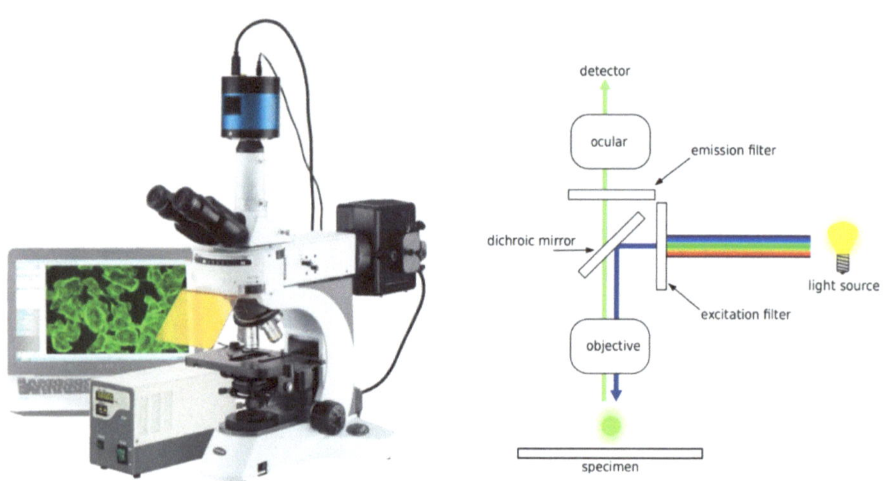

Virtual learning aid

Confocal microscope

A **confocal microscope** enhances the clarity of specimen images, **overcoming the blurriness found in conventional microscopes**. It achieves this by excluding most of the light that doesn't originate from the microscope's focal plane. The result is an image with reduced haze and improved contrast, providing a thin cross-section view of the specimen. This not only allows better observation of fine details but **also enables the construction of three-dimensional (3D) reconstructions** by assembling a series of thin slices along the vertical axis.

In the context of microscopy, the term "**confocal**" refers to the **same focus** shared by the object and its image. The confocal microscope accomplishes this by filtering out out-of-focus light from both above and below the focal point in the specimen. Typically, fluorescence microscopes capture signals from the entire thickness of the specimen, leading to an unfocused view for the observer. However, the confocal microscope overcomes this limitation by incorporating

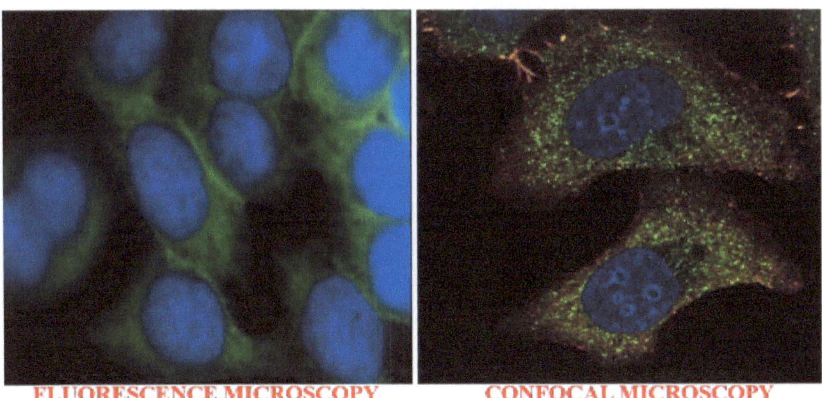

a **confocal pinhole** as a spatial filter in front of the image plane. This pinhole **permits only the in-focus portion of light** to be imaged, effectively **eliminating the out-of-focus information**. The process involves illuminating a fluorescently stained specimen with a laser beam to facilitate visualization.

Virtual learning aid

Electron microscope

Electron microscope shares fundamental principles with light microscopy, but with one major difference: it **employs electromagnetic lenses to focus a**

high-velocity electron beam instead of visible light. This similarity extends to the relationship between resolution limits and the wavelength of illuminating radiation, which applies to both electron and light beams. Due to the short wavelength of electrons, electron microscopes boast exceptionally high resolving power.

There are **two** primary types of electron microscopes: **transmission electron microscope (TEM)** and **scanning electron microscope (SEM)**. The transmission electron microscope forms images by transmitting electrons through the specimen under examination. In contrast, the scanning electron microscope operates differently, producing images from electrons that have been deflected from the outer surface of the specimen.

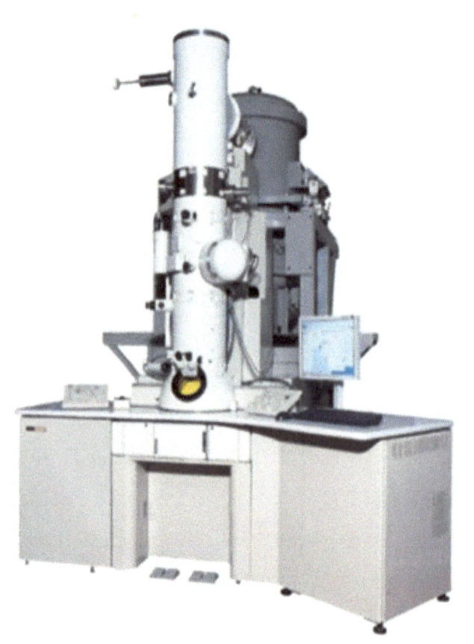

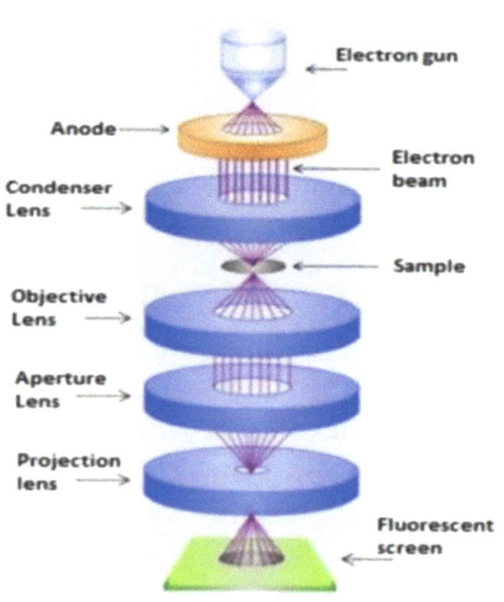

Virtual learning aid

Note

Electron microscopes consist of a tall, hollow cylindrical column where an electron beam passes. At the top of the column, there's a cathode—a tungsten wire filament heated to provide a source of electrons. These electrons are then accelerated into a fine beam by applying a high voltage between the cathode and anode. Similar to how glass lenses focus light rays in light microscopes, electromagnetic lenses focus the beam of negatively charged electrons. The strength of the magnetic lenses can be controlled by adjusting the current provided to them. To avoid electron scattering due to collisions with air molecules, the column is evacuated to create a vacuum through which the electrons travel.

Comparison between Light microscope and Electron microscope.

Feature	Light Microscope	Electron Microscope (TEM)
Highest practical magnification	About 1,000-1,500	Over 100,000
Best resolution	0.2 μm	0.5 nm
Radiation source	Visible light	Electron beam
Medium of travel	Air	High vacuum
Type of lens	Glass	Electromagnet
Source of contrast	Differential light absorption	Scattering of electrons
Specimen mount	Glass slide	Metal grid (usually copper)

Transmission electron microscope

In light microscopy, differences in how specimens absorb light, mainly due to staining, create visible variations in the image. However, in transmission electron microscopy (TEM), **a condenser lens focuses an electron beam onto the specimen, with minimal absorption by the sample**. Instead, electrons are scattered by the atoms in the specimen, and image formation depends on this **differential scattering**. When no specimen is present, the screen is uniformly bright due to an evenly illuminated electron beam. But when a specimen is introduced, some electrons are scattered away, leading to areas of reduced electron flux and creating contrast in the image.

The interaction between the incoming fast electrons and the atomic nuclei causes **elastic scattering**, with no energy loss. In contrast, inelastic scattering occurs when fast electrons interact with atomic electrons, transferring some energy. The image is formed by focusing the unscattered electrons, similar to a light microscope, and can be observed on a **phosphorescent screen** or recorded using **photographic plates** or **digital cameras**. To maintain the electron beam quality, the entire tube between the electron source and detector is kept under an ultrahigh vacuum, as electrons are absorbed by atoms in the air.

The amount of scattering at any point in the specimen depends on its density and overall thickness, making contrast relatively independent of other specimen properties. TEM specimens are typically prepared with uniform thickness, reducing contrast from thickness variations. Their atomic number remains constant, leading to low overall contrast and a featureless appearance. **To enhance scattering contrast, specimens are stained with** electron-dense materials, such as salts of **heavy metals like uranium and lead**. Immersing the sample in such solutions before or after slicing reveals different cellular components with varying degrees of contrast, with darker areas indicating reduced electron transmission due to thickness or staining.

Virtual learning aid

Scanning electron microscope

The **scanning electron microscope (SEM)** is a powerful tool used to visualize the surfaces of specimens. First, the sample is fixed, dried, and coated with a thin layer of a heavy metal like **gold** or **gold-palladium mixture**. Next, a narrow beam of electrons scans the specimen, exciting its molecules, which then **release secondary electrons**. These electrons are detected, generating a detailed image of the specimen's surface. The contrast in the image arises from the varying amounts of secondary electrons produced by different parts of the specimen upon electron beam impact. The SEM provides a **resolution of about 10 nm**, slightly less than the transmission electron microscope, but its **images are three-dimensional** compared to the two-dimensional images of the latter.

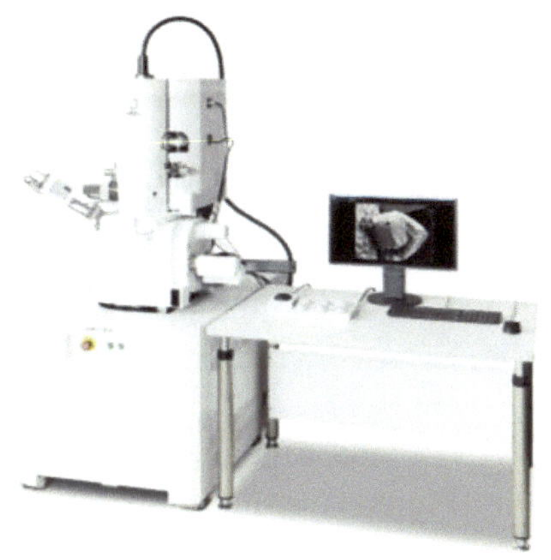

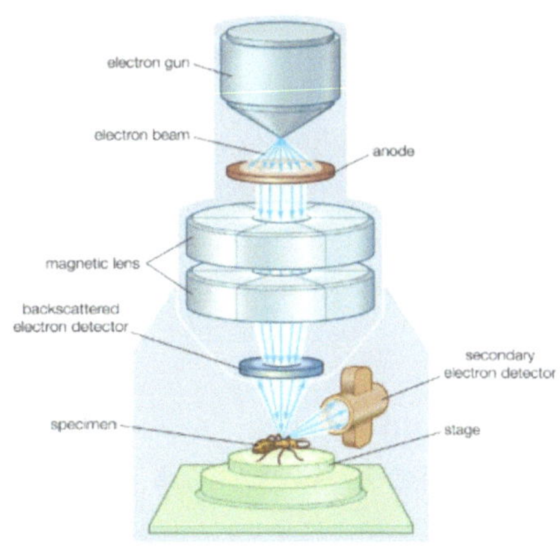

Virtual learning aid

Sample preparation for microscope

Sample preparation for light microscopy involves several essential steps to ensure proper visualization and analysis of the specimen. These steps include **sample collection, fixation, washing, dehydration, embedding, sectioning, staining**, and **mounting**.

1. **Sample Collection**: The specimen can be a whole organism, organ, squashed tissue, cells suspended in fluid, or sections of a larger sample.

2. **Fixation**: Fixatives are used to stabilize the sample's structure, preserving it in a life-like state. This process makes the cells permeable to staining reagents and locks their macromolecules in position. Fixation can be achieved through physical methods like **heating or freezing**, but chemical fixation is more commonly used. **Chemical fixatives**, such as **formaldehyde** and **glutaraldehyde**, cross-link proteins, and nucleic acids, effectively preserve the specimen's structural appearance.

149

3. **Washing**: After fixation, excess fixative is washed out of the tissue to remove any residual chemicals.

4. **Dehydration**: Water is removed from the specimen by gradually immersing it in increasing concentrations of ethanol or acetone.

5. **Embedding**: The specimen is hardened, making it suitable for sectioning. Embedding materials, like paraffin wax or resin, are used for this purpose.

6. **Sectioning**: The hardened specimen is sliced into thin sections using a microtome or cryotome, allowing visualization of internal structures.

7. **Staining**: Staining is crucial for enhancing contrast in light microscopy. Dyes, with specific functional groups, are used to color the specimens. **Basic dyes**, such as **crystal violet** and **methylene blue**, react with negative charge groups, while **acidic dyes**, like **eosin** and acid **fuchsin**, react with positively charged groups. Stains bind to specific molecules in the specimens, making certain parts of the sample visible under the microscope.

8. **Mounting**: The fixed and stained specimens are mounted between slides and cover slips using adhesives. This prepares the slides for storage or observation. Different mounting methods can be used, including wet or dry mount, temporary or permanent mount, and hot or cold mount.

Sample preparation for TEM

Transmission Electron Microscopy (TEM) requires ultra-thin cross-sections of samples to enable electrons to pass through them. The cutting process is performed using an instrument called an **ultramicrotome** after the sample has been fixed and dehydrated. The contrast in TEM relies on the atomic number of the atoms present in the specimen; **higher atomic numbers** lead to **increased electron scattering** and **greater contrast**. Since biological tissues primarily consist of atoms with low atomic numbers (such as carbon, oxygen, nitrogen, and hydrogen), heavy metal salts are employed as stains in electron microscopy. **Colors are not observed in electron microscopes**.

To enhance contrast and visibility, electron stains can produce positive or negative effects. **Positive contrast** staining involves using **heavy metal salts** to increase the density of specific biological structures, thus creating differential contrast. This staining method utilizes stains like **uranyl acetate** and **lead citrate**, which have an affinity for nucleic acids and certain proteins. The exact mechanism of these stains is not fully understood, but it is known that uranyl ions bind strongly to phosphate and amino groups, leading to significant staining. Conversely, **negative staining** focuses on **staining the background** surrounding the specimen, leaving the actual sample relatively untouched. Negative stains are suitable **for visualizing whole biological structures** like viruses, bacteria, and cellular organelles. Common negative stains include **uranyl acetate** and **phosphotungstic acid**.

Positive staining, although useful in providing detailed information, can sometimes obscure fine details when specimens are coated with stain. In contrast, **negative staining preserves the fine structure and morphology in the specimens**, as the stain forms a mold around them. In electron micrographs, negatively stained samples appear as light regions against a dark background due to the action of heavy metal salts like uranium, tungsten, or molybdenum. While positive staining is employed in sectioned materials, negative staining is utilized to enhance the visualization of intact biological structures, offering a valuable tool in electron microscopy research.

Freeze-fracture electron microscopy

Freeze-fracture electron microscopy is a valuable technique **for visualizing cell membrane features**. To avoid distortions caused by ice crystal formation during freezing, cells are rapidly **frozen in liquid nitrogen (-196°C)** with the **aid of a cryoprotectant like glycerol**. The frozen cells are then fractured, revealing the interior of the lipid bilayer and its embedded proteins. A replica or cast is made of the fractured surface through a **two-step** process: **oblique shadowing** with heavy metal (e.g., platinum) deposition, followed by **carbon backing**. This creates variations in thickness on the replica, representing the topographical features of the frozen, fractured surface.

Once the freeze-fracture replica is created, the sample is brought back to room temperature, and the **biological material is removed** using cleaning agents like sodium **hypochlorite or chromic acid**. After thorough washing, the replica pieces are mounted on grids for examination using a **transmission electron microscope**. The **four essential steps** in making a freeze-fracture replica are **rapid freezing**, **specimen fracturing**, **vacuum-deposition** on platinum and carbon for **replica creation**, and finally, **cleaning** to eliminate all biological material. Optionally, an etching step involving vacuum sublimation in ice can be inserted between fracturing and replica-making, referred to as "**freeze etching**."

Freeze-fracture electron microscopy provides a clear view of cell membrane structures. By rapidly freezing cells in liquid nitrogen with glycerol as a cryoprotectant, ice crystal formation is prevented. Fracturing the frozen cells exposes the lipid bilayer and its embedded proteins, which are then replicated by oblique shadowing with heavy metal (platinum) deposition, followed by carbon backing. Cleaning the replica thoroughly and optionally employing freeze etching ensures the removal of biological material before examination in the transmission electron microscope.

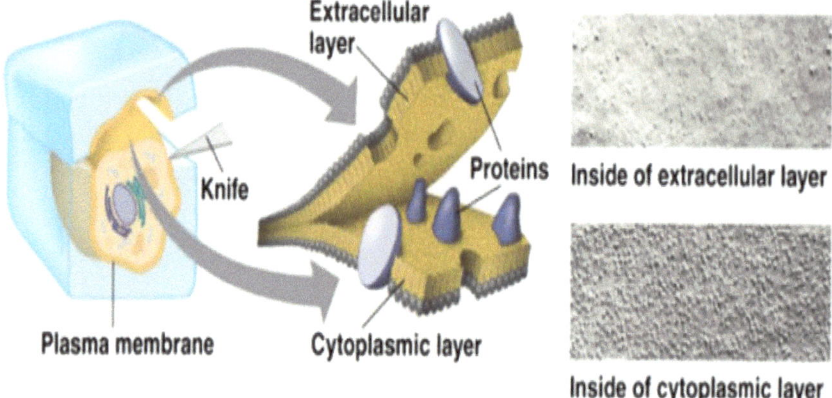

Virtual learning aid

Cryo-electron microscopy (Resolution revolution)

Cryo-electron microscopy (Cryo-EM) is a powerful technique **for determining the high-resolution structures of biomolecules in solution**. Unlike traditional transmission electron microscopy (TEM), Cryo-EM uses frozen biological samples **at cryogenic temperatures** to avoid sample damage caused by high-energy electron beams and vacuum conditions. By rapidly freezing the samples using **liquid nitrogen-cooled ethane**, the water molecules form a disordered glass-like state, preventing crystalline ice formation and electron diffraction. This **vitrification process preserves the natural shape and organization of the biomolecules**.

In **Cryo-EM**, a focused beam of low-energy electrons interacts with the frozen sample, **producing 2D snapshots from different angles**. The challenge lies in the fuzziness of these images due to the low-energy electron beam. To overcome this, a sophisticated image-analysis algorithm merges and processes thousands of randomly oriented molecule images, generating much sharper 2D images. These images are then used to calculate the spatial relationships between the molecule groups, **ultimately assembling a high-resolution 3D image in the biomolecule**.

The advantages of Cryo-EM lie in its ability to maintain the integrity in biological samples and provide detailed structural information. By avoiding the destructive effects of high-energy electron beams and preventing the formation of ordered ice crystals, Cryo-EM offers a valuable approach to visualize the intricate structures of biomolecules in their native state.

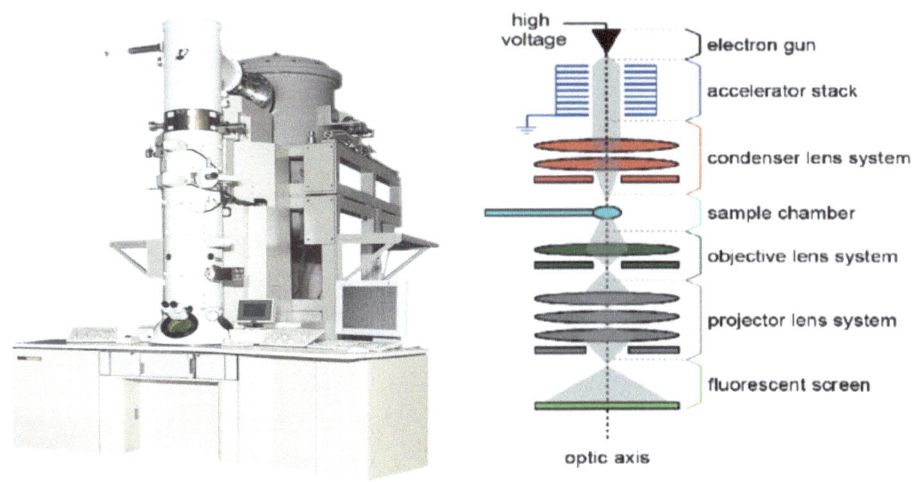

Cryo-EM Vs X-ray crystallography

X-ray crystallography is a powerful method used **to obtain high-resolution structures of biomolecules by analyzing X-ray diffraction patterns** from a crystal. By striking a crystal with X-rays, the resulting diffraction pattern reveals the arrangement of atoms in a three-dimensional picture of the electron density within the crystal. However, not all biomolecules can form crystals, and in some cases, the crystallization process may alter the molecule's natural structure, limiting its representativeness. In contrast, **Cryo-electron microscopy (Cryo-EM)** does not require crystallization and allows scientists **to visualize biomolecules in their native state**, capturing their movements and **interactions during their functional processes**.

Cryo-EM Vs NMR

Nuclear Magnetic Resonance (NMR) provides intricate **structural information** and is valuable **for studying conformational changes** and **binding events of small biological molecules**. However, NMR has limitations when dealing with larger molecular masses (<30 kDa) and membrane-bound molecules. For studying high molecular mass proteins, membrane-bound receptors, and biomolecular complexes, Cryo-Electron Microscopy (cryo-EM) emerges as the preferred technique.

Scanning probe microscopes

Scanning Probe Microscopes (SPMs) belong to a group of powerful microscopes used **to examine surface topography at the atomic level**, with **resolutions reaching the nanometer scale**. These microscopes work by employing a physical probe to scan the surface of a sample, measuring the interactions between the probe tip and the sample's surface. The gathered data is then used to generate a detailed image of the sample's surface. Different types of SPMs utilize various interactions between the probe tip and the sample, resulting in distinct imaging capabilities. **Two major SPM** types are the **Scanning Tunneling Microscope (STM)** and the **Atomic Force Microscope (AFM)**. **AFMs measure interactive forces between the probe tip and the sample**, while **STMs measure the tunneling current flowing**

between the conducting sample and the probe tip. Notably, STM requires a conductive sample for surface imaging, making it unsuitable for most biological samples due to its poor electron conductivity. In contrast, AFMs do not require conducting samples.

The **Scanning Tunneling Microscope (STM)** holds a pivotal place in the family of SPMs, being the first of its kind, invented **by Binning and Rohrer in 1981**. Its operating principle relies on the phenomenon of electron tunneling, where electrons can tunnel through a narrow potential barrier between a metal tip and a conducting sample in the presence of an external electric field. By connecting the probe tip and the sample to a voltage source and positioning the tip close to the sample surface, electrons start to tunnel across the gap between the tip and the surface. The magnitude of the tunneling current depends on the distance between the tip and the sample surface, enabling the STM to achieve exceptionally high spatial resolution. This exponential dependence of tunneling current on the tip-sample distance allows for detailed imaging of atomic-scale features on conductive surfaces.

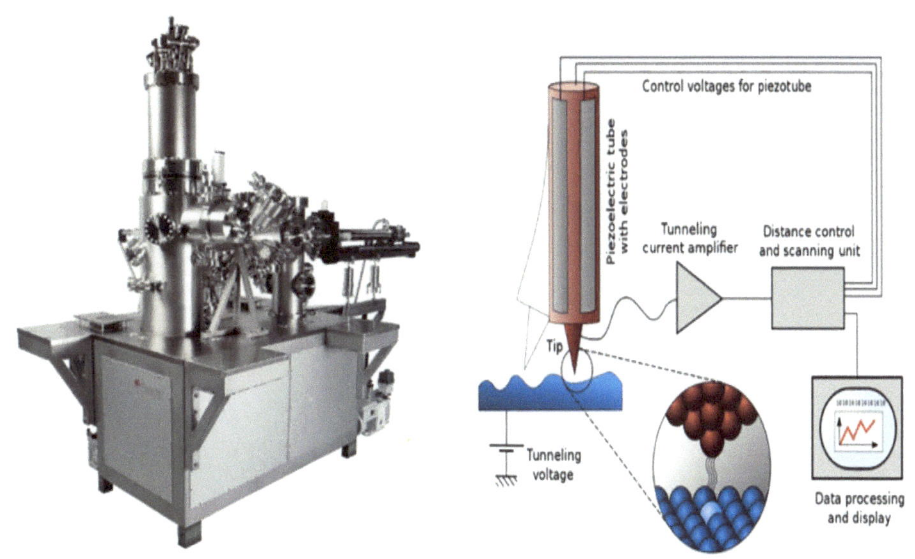

Atomic-force microscope

The **Atomic Force Microscope (AFM)** operates by measuring **interactive forces between a probe tip and a sample surface**. These forces, such as capillary, electrostatic, and van der Waals forces, can be short or long-range, and either attractive or repulsive. As the tip scans the surface, it encounters these interactions, leading to movements that cause the cantilever to deflect. Attractive forces make the cantilever deflect towards the surface, while repulsive forces push it away. By measuring this deflection, AFM can map out the surface's topographical features. The AFM uses **three imaging methods**: **contact mode**, **tapping (intermittent) mode**, and **non-contact mode**. In contact mode, the tip directly touches the sample surface, primarily experiencing repulsive forces. Tapping mode involves intermittent contact and uses the cantilever's resonant frequency to maintain a constant tip-sample interaction during scanning. Non-contact mode keeps the probe tip above the surface, experiencing predominantly attractive forces.

Contact mode in AFM, while effective, has the drawback of direct mechanical interaction between the probe tip and the sample surface. Tapping mode overcomes this by using intermittent contact, reducing the wear on the tip and the sample. The oscillation of the cantilever at its resonant frequency ensures a constant interaction, enabling surface imaging. Non-contact mode, on the other hand, operates without the probe tip physically touching the surface. Instead, it oscillates above the surface, primarily experiencing attractive forces. Each mode offers distinct advantages for imaging, catering to different sample types and research needs.

Virtual learning aid

<h1 align="center">Chapter 9</h1>

<h1 align="center">Immunotechniques</h1>

Introduction

Antigen-antibody interactions are **highly specific**, resembling the way enzymes associate with their substrates. These interactions involve **weak** and **reversible non-covalent forces**, including van der Waals forces, electrostatic forces, H-bonding, and hydrophobic forces. The smallest unit of an antigen capable of binding with antibodies is called an **antigenic determinant or epitope**. On the antibody molecule, the corresponding region that binds to the epitope is known as the **paratope**. The **valence of an antigen** refers to the number of epitopes present on its surface, determining how many antibody molecules can bind to it simultaneously. When an antigen has one epitope, it is **monovalent**, while most antigens are **polyvalent**, having multiple copies of the same epitope. Immunoprecipitation occurs when a soluble antibody interacts with a soluble antigen, forming an insoluble complex. This reaction depends on both the antibody and antigen being bivalent or polyvalent. Monovalent Fab fragments cannot initiate a precipitation reaction, and only when the resulting cross-linked complex becomes large enough, it precipitates out of the solution. Immunoprecipitation can be performed in a solution or a gel matrix.

Immunoprecipitation reaction in solution

Precipitation reaches its maximum when the antigen and antibody are present in optimal proportions, leading to the formation of insoluble antigen-antibody complexes within a specific concentration range known as **zone of equivalence**. This range ensures that the complexes formed are large enough to be precipitated. Conversely, outside this zone, when either antigen or antibody is present in excess, smaller soluble complexes are formed. Consequently, when increasing antigen concentrations are introduced to tubes containing a constant antibody concentration, varying amounts of precipitate are formed.

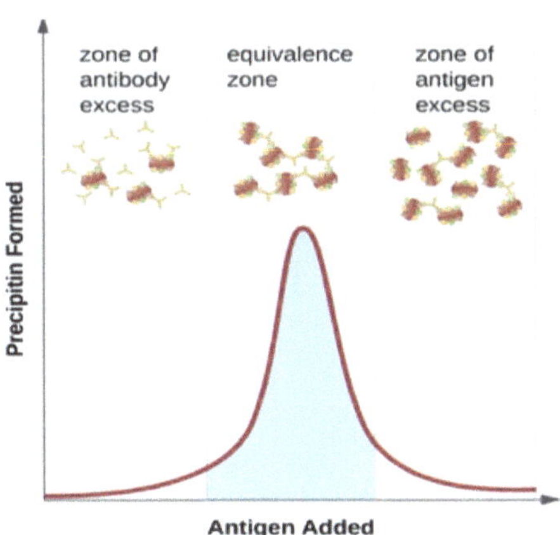

Immunoprecipitation reactions in gels

Immunoprecipitation reactions conducted in **agar gels** are termed **immunodiffusion reactions**. In these reactions, antigen and antibody diffuse towards each other in the gel, resulting in the formation of a visible line in precipitation called the **precipitin line**. This line appears in the region where antigen and antibody are present in equivalent amounts, while no visible precipitate forms in areas of antibody excess or antigen excess. There are **two types** of immunodiffusion techniques, namely **radial immunodiffusion** and **double immunodiffusion**, which can be utilized **to determine the relative concentrations of antibodies or antigens and to identify specific antigens**.

Virtual learning aid

Radial immunodiffusion (Mancini method)

A simple **quantitative assay** is used **to determine the relative concentration of an antigen**. In this method, the antigen sample is placed in a well and allowed to diffuse into an agar gel containing a uniformly distributed antibody. As the antigen and antibody react at the region of equivalence, a **precipitation ring** forms around the well. The **diameter of this ring is directly proportional to the logarithm of the antigen concentration**, with the amount of antibody being constant. By running different concentrations of a standard antigen on the gel and measuring the diameters of their precipitin rings, a calibration graph is constructed. This graph allows researchers to estimate the antigen concentrations of unknown samples by measuring the diameter of their precipitin rings and extrapolating the value on the calibration graph.

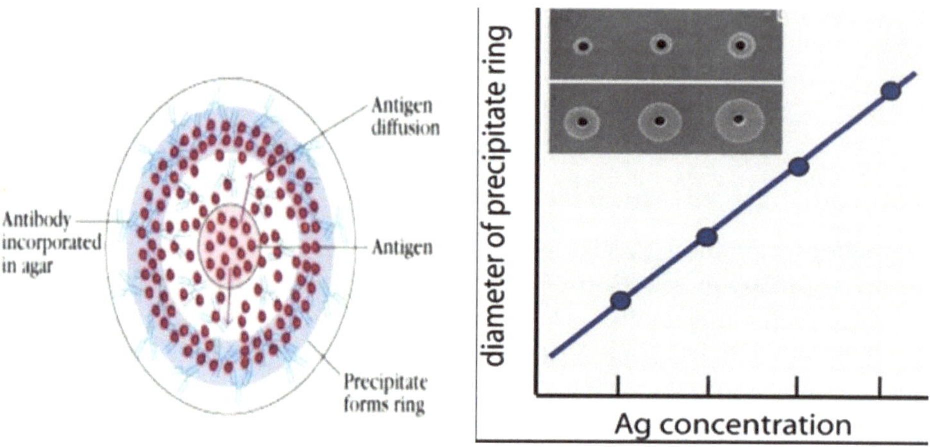

Reagents

1. Agarose powder
2. Buffered saline solution (e.g., Phosphate-buffered saline, PBS)
3. Specific antibody against the target antigen
4. Antigen standard of known concentration
5. Staining dye (e.g., Coomassie Brilliant Blue)
6. Running buffer (Tris-Glycine buffer)

Equipment

1. Petri dishes or plates
2. Glass plates or slides
3. Micropipettes and pipette tips
4. Water bath
5. Refrigerator
6. Electrophoresis chamber
7. Power supply for electrophoresis
8. Image analysis software (optional)

Protocol

1. **Preparation of Gel Plate**: Prepare a **1% agarose gel** by dissolving 1g agarose powder in 100 mL of buffered saline solution (PBS) in a flask. Heat the flask containing the mixture in a water bath until the agarose is completely dissolved. Allow the agarose solution to cool down to about **50-60°C**. Pour the agarose solution into a clean and dry **Petri dish** (or any flat-bottomed container) to a depth of about **3-5 mm**. Let the agarose gel solidify at room temperature or place it in the refrigerator for quicker solidification.

2. **Setting Up the Immunodiffusion**: Using a **sterile punch**, create evenly spaced wells in the agarose gel. Add a specific antibody solution into each well using a micropipette. Ensure that the antibody concentration is uniform in all wells. Prepare a separate well for the antigen standard of known concentration.

3. **Diffusion and Precipitation**: Add the sample containing the antigen interest to the antigen well. Cover the Petri dish to prevent contamination and evaporation. Incubate the Petri dish at an appropriate temperature (usually at room temperature) to allow antigen-antibody interaction and diffusion.

4. **Electrophoresis**: If necessary, perform electrophoresis to improve the antigen-antibody interaction and reduce the diffusion time. Set up the electrophoresis chamber with a **Tris-Glycine buffer**. Place the gel plate with the wells in the electrophoresis chamber. Connect the gel plate to the power supply and run the electrophoresis at low voltage (e.g., **20-30V**) for a short period (usually **10-15 minutes**).

5. **Staining and Analysis**: After the completion of diffusion, remove the gel plate from the Petri dish. Stain the gel using **Coomassie Brilliant Blue** or any suitable staining dye. Follow the staining protocol for the specific dye used. Destain the gel to remove excess stain and enhance the visibility of the precipitin rings. Analyze the gel by measuring the diameters of the precipitin rings using a ruler or image analysis software. Compare the diameter of the antigen precipitin rings with the standard curve generated from known antigen concentrations to determine the concentration of the antigen in the sample.

Ouchterlony double immunodiffusion

Ouchterlony double immunodiffusion is a technique used in immunology **to detect, identify**, and **quantify antibodies and antigens**. In this method, **both the antigen and antibody diffuse radially** from separate wells toward each other, creating a concentration gradient. The rate of diffusion depends on the concentration of antigens and antibodies in the well and their sizes. When they meet at equivalence, a precipitin line forms. This technique allows us to analyze the antigenic relationship between two antigens. Different precipitation line patterns are observed depending on whether the antigens share all, some, or no antigenic epitopes:

1. **Identity**: Occurs when two antigens share identical epitopes, resulting in a **single continuous precipitin line**.

2. **Non-Identity**: Occurs when two antigens are unrelated and do not share any common epitopes. The antisera form **independent precipitin lines** with each antigen and the two lines cross.

3. **Partial Identity**: Occurs when two antigens share some epitopes but have unique epitopes as well. The antiserum forms **a line of identity with the common epitopes and a curved spur** with the unique epitope.

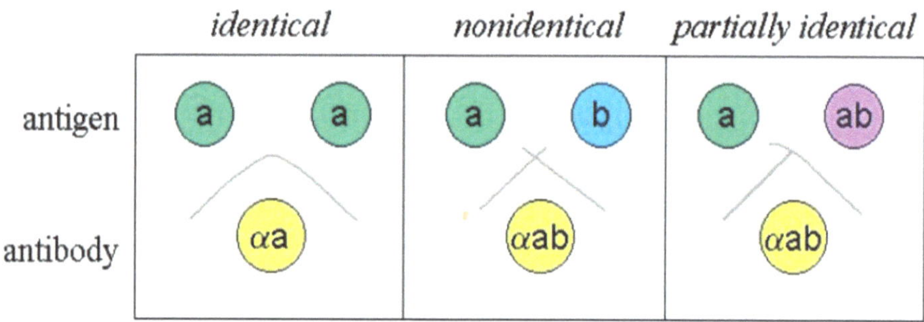

Virtual learning aid

Materials and Equipment

1. Agarose powder
2. Electrophoresis-grade buffer (Tris-acetate-EDTA or Tris-borate-EDTA)
3. Antigen solution (purified or extracted from the sample of interest)
4. Antibody solution (specific to the antigen of interest)
5. Microcentrifuge tubes or test tubes
6. Micropipettes and pipette tips
7. Glass slides
8. Aluminium foil
9. Petri dishes
10. Water bath or hot plate
11. Refrigerator

Protocol

1. **Preparation of Agarose Gel**: Weigh the appropriate amount of agarose powder according to the desired gel concentration and add it to a flask containing the electrophoresis buffer. Heat the mixture using a water bath or hot plate while stirring until the agarose is completely dissolved. Pour the agarose solution into a clean and sterile petri dish to create the gel bed. Allow it to solidify at room temperature.

2. **Preparing Wells**: Use a micropipette to create multiple wells (usually 5-6) on the agarose gel bed while it is still soft. Arrange the wells in a circular pattern, leaving enough space between them. Allow the gel to solidify completely and remove any excess buffer from the petri dish.

3. **Preparing Samples**: In separate microcentrifuge tubes, prepare the antigen and antibody solutions by diluting them appropriately in the buffer to a desired concentration. Mix the solutions gently to ensure homogeneity.

4. **Loading Samples**: Carefully load the antigen solution into the wells on the agarose gel bed using a micropipette. In the adjacent wells, load the antibody solution in a similar manner. In one well, load only the buffer as a negative control.

5. **Diffusion and Incubation**: Cover the petri dish with a clean glass slide and wrap it with Aluminium foil to create a humid chamber, preventing evaporation. Incubate the petri dish in a refrigerator at an appropriate temperature (usually between 4°C to 8°C) for 24 to 48 hours, allowing the diffusion of antigens and antibodies to form immunoprecipitation lines.

6. **Result Interpretation**: After the incubation period, carefully remove the glass slide and the Aluminium foil covering. Examine the gel for the presence in precipitin lines that form where the antigen and antibody have interacted. Measure the distance between the centers of the precipitin lines to estimate the relative size of the antigen-antibody complexes. Compare the results to the negative control well to confirm specific antigen-antibody reactions.

Note

Zeta potential refers to the electrical charge present on the surfaces of certain particulate antigens. For example, erythrocytes possess a net negative charge due to the presence of sialic acid. When these charged particles are suspended in a saline solution, they create an electrical potential known as the zeta potential. This prevents the particles from coming too close to each other, making it challenging for antibodies to agglutinate or clump them together.

The **Ouchterlony method** is utilized **to estimate the relative concentration of antigens**. When an antigen has a higher concentration, the equivalent zone forms a little farther away from the antigen well. Conversely, when an antigen has a lower concentration, the equivalent zone forms a little closer to the antigen well. This method helps researchers analyze and compare the levels of different antigens based on the positioning of their respective equivalent zones.

Agglutination reactions

Agglutination is a visible clumping of a particulate antigen when mixed with specific antibodies. These **antibodies**, known as **agglutinins**, can cause this clumping reaction. The term "agglutinin" encompasses any antibody that agglutinates particulate antigens, while **"hemagglutination"** is used when the antigen is an erythrocyte. Among antibodies, **IgM** is particularly effective as an agglutinin due to its **high valence**, allowing it to cross-link multiple antigens. Agglutination reactions are similar to precipitation reactions, and both can be inhibited by **an excess of antibodies**, known as the **prozone effect**, or by **an excess of antigens**, resulting in the **postzone effect**. **Agglutination tests** can be either **direct (active)**, where the antigen is an integral part of a cell's surface, or **indirect (passive)**, using soluble antigens coated on insoluble particles like cells.

Virtual learning aid

Scan It!

The agglutination test exclusively applies to particulate antigens. Nevertheless, it is possible to employ passive agglutination by coating insoluble particles, such as cells, with soluble antigens like viral antigens, polysaccharides, or haptens, enabling the detection of antibodies against the soluble antigens. Understanding agglutination is crucial in diagnostic and research settings, as it aids in detecting specific antigens or antibodies. IgM's potent agglutinating ability is particularly valuable in diagnostic assays.

Coombs test

When antibodies attach to red blood cells (erythrocytes), they don't always cause clumping (agglutination). This can happen due to the ratio of antigens to antibodies (either an excess of antigens or antibodies). Additionally, the zeta potential on the erythrocytes can prevent effective cross-linking of the cells. To identify the presence of non-agglutinating antibodies on erythrocytes, a second antibody directed against the antibodies attached to the erythrocytes is added. This secondary antibody, known as anti-immunoglobulin, can then cause agglutination by cross-linking the erythrocytes.

The **Coombs test (anti-immunoglobulin test)** is based on **two** important principles. First, **antibodies from one species** (e.g., human) become **immunogenic when introduced into another species** (e.g., rabbit), leading to the production of antibodies against those antibodies. Second, **many anti-immunoglobulins** (e.g., rabbit anti-human antibodies) **bind to antigenic determinants present on the Fc portion of the antibody**, leaving the **Fab portions free to react with antigens**. For example, when human IgG antibodies are attached to their respective epitopes on erythrocytes, the addition of rabbit antibodies to human IgG will result in the rabbit antibodies binding to the Fc portions of the human antibodies on the erythrocytes via their Fab portions. This binding not only links the human antibodies on the erythrocytes but also forms cross-links between them.

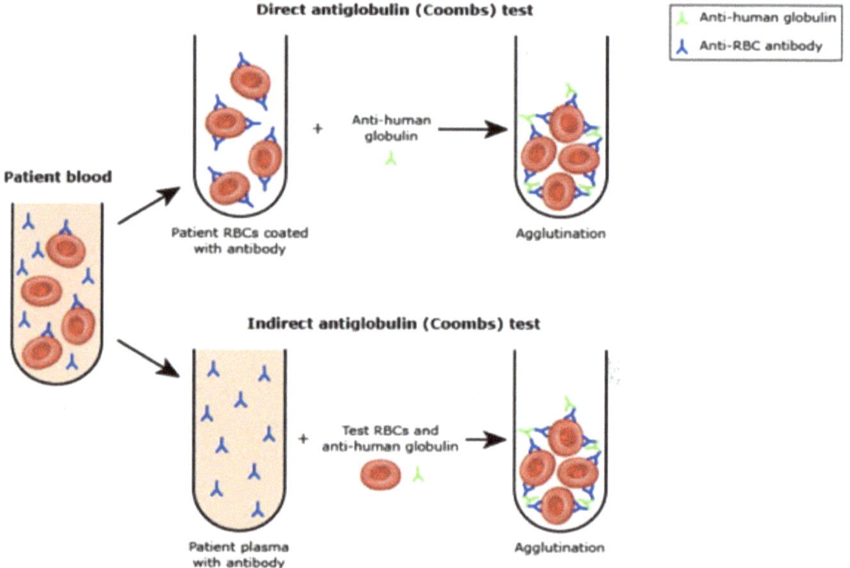

The Coombs test comes in two versions: direct and indirect. Both are based on the same principle. In the direct test, anti-immunoglobulins are added to suspected erythrocytes with bound antibodies. Meanwhile, the indirect test detects specific antibodies to erythrocyte antigens in the serum.

Virtual learning aid

Scan It!

Materials

1. Patient blood sample (EDTA-anticoagulated)
2. Normal saline (0.9% NaCl) solution
3. Coombs test kit (commercially available)
4. Anti-human globulin (AHG) reagent (monospecific or polyspecific)
5. Microcentrifuge tubes
6. Centrifuge
7. Pipettes (single and multichannel)
8. Graduated cylinders
9. Timer or stopwatch
10. Incubator (at 37°C)

Equipment

1. Spectrophotometer (for optional quantification)

Protocol

1. **Sample Collection and Preparation**: Obtain a patient's blood sample in an **EDTA-anticoagulated tube** using aseptic techniques. Label the tube appropriately with the patient's details for proper identification.

2. **Washing the Red Blood Cells (RBCs)**: Gently mix the blood sample to ensure even distribution of RBCs. Carefully transfer **2-3 mL** of the well-mixed blood into a microcentrifuge tube. Add an equal volume of **normal saline (0.9% NaCl)** to the microcentrifuge tube containing the blood sample. Mix the contents of the tube thoroughly and centrifuge at **1,500-2,000 rpm** for **2-3 minutes**.

3. **Decanting the Supernatant**: After centrifugation, gently decant the supernatant from the tube, taking care not to disturb the RBC pellet.

4. **Coombs Test (Direct Antiglobulin Test)**: Add a drop of **anti-human globulin** (AHG) reagent (monospecific or polyspecific) to the RBC pellet in the microcentrifuge tube. Mix gently to ensure proper coating of the RBCs with the AHG reagent. Incubate the tube in an incubator set at **37°C** for **15-30** minutes.

5. **Centrifugation and Observation**: Centrifuge the tube again at **1,500-2,000 rpm** for **2-3 minutes**. Observe the RBC pellet for agglutination under good lighting conditions. Agglutination indicates a positive reaction.

6. **Optional Quantification (for positive reactions)**: In the case of a positive reaction, the degree of agglutination can be quantified using a spectrophotometer to measure the absorbance at specific wavelengths.

Note

The Coombs test, or Direct Antiglobulin Test (DAT), is an essential diagnostic tool used to identify immune-mediated hemolytic conditions. By detecting antibodies or complement proteins on the surface of red blood cells, the Coombs test aids in diagnosing various blood disorders.

Immunoassays

Immunoassays are powerful biochemical methods used **to measure the presence or concentration of a specific analyte (known as an antigen) in a solution**. These assays are **based on the specific interactions between antibodies and antigens**. When an antigen is present in the solution, it binds to its corresponding antibody, forming an immune complex. Immunoassays are preferred for their exceptional specificity, sensitivity, and ability to detect low concentrations of analytes. To facilitate detection, labels such as radioisotopes or enzymes are attached to either the analyte or the antibody. There are **two** main types of immunoassays: **competitive** and **noncompetitive**. In **competitive immunoassays**, a **labeled antigen competes with an unlabeled antigen of interest** for a limited amount of specific antibodies. On the other hand, **noncompetitive immunoassays** use **a labeled antibody to detect the antigen of interest**, requiring an excess of labeled antibodies.

Competitive immunoassays, like traditional radioimmunoassays (RIA), necessitate only a small amount of antibodies and rely on the competition between labeled and unlabeled antigens for binding to the specific antibody. Noncompetitive immunoassays, in contrast, utilize an abundance of labeled antibodies to detect the antigen of interest directly. To optimize sensitivity, competitive assays benefit from reduced antibody amounts, while non-competitive assays achieve maximal sensitivity by increasing the antibody concentration.

Radioimmunoassay

Radioimmunoassay (RIA) is a highly **sensitive** and **specific** method used **to quantify the concentration of a specific antigen in a sample**. It was first developed in **1959 by Yalow and Berson**, who used **radiolabeled insulin** to measure insulin levels in human plasma. The basic principle of **RIA** involves competitive binding, **where a radiolabeled antigen competes with the unlabeled antigen in the sample for a fixed number of antibodies**. After incubating the sample antigen with the antibody, the radiolabeled antigen is introduced, displacing the sample antigen from the antibody. The more antigens present in the sample, the less radiolabeled antigen binds to the antibody. To calibrate the measurements, known amounts of unlabeled antigens are added to tubes containing fixed amounts of antibodies and radiolabeled antigens. By measuring counts from these tubes, a standard curve is generated, enabling the determination of antigen concentration in an unknown sample by referencing the curve.

To perform a radioimmunoassay, one needs a sample having antigen of interest, a corresponding antibody, and a radiolabeled version of the antigen. By allowing the antigen and antibody to bind and subsequently introducing the radiolabeled antigen, researchers can quantify the sample antigen based on the competitive binding between labeled and unlabeled antigens. The higher the concentration of the unlabeled antigen in the sample, the better it competes for the binding sites on the antibody, leading to fewer bound radiolabeled antigens and lower **counts per minute (CPM)**. This competition allows for the generation of a standard curve using tubes with varying known concentrations of the unlabeled antigen. Each experiment requires its standard curve to accurately determine the antigen concentration in unknown samples.

Enzyme-linked immunosorbent assay

Enzyme immunoassays (EIA) are **diagnostic techniques** that utilize **enzymatic reactions to detect immune reactions**. In **1971, Engvall and Perlmann, as well as Weemen and Schuurs independently**, introduced the concept of using enzyme-linked agents. The most widely used type of enzyme immunoassay is the **enzyme-linked immunosorbent assay (ELISA)**, which is akin to radioimmunoassays (RIA) but employs enzymes instead of radioactive labels for detection. In ELISA, an enzyme linked to an antibody triggers a reaction with a colorless substrate, resulting in the generation of a colored product. This substrate is termed a **chromogenic substrate**. Several enzymes, such as **alkaline phosphatase, horseradish peroxidase, urease**, and **beta-galactosidase**, have been employed in ELISA. Additionally, detection can also occur using fluorescently labeled antibodies, leading to a **fluorescence-linked immunosorbent assay (FLISA)**.

List of Enzymes, their sources, and the reactions they catalyze in ELISA.

Enzyme	Source	Reaction Catalyzed
Peroxidase	Horseradish	H_2O_2 + Oxidisable substrate $\rightarrow$ Oxidized product + $2H_2O$
Alkaline phosphatase	Calf intestine	$R\text{-}O\text{-}P_1 + H_2O \rightarrow R\text{-}OH + P_1$
B-Galactosidase	E. coli	B-D-Galactoside + $H_2O \rightarrow$ Galactose + Alcohol
Urease	Jack bean	$(NH_2)_2\,CO + 3H_2O \rightarrow CO_2 + 2NH_4OH$

ELISA, or Enzyme-Linked Immunosorbent Assay, has been adapted into different variants for either qualitative detection or quantitative measurement for target molecules. This division results in **two main types** of ELISA: **qualitative**, which gives a straightforward **positive or negative outcome**, and **quantitative**, which provides information on the **concentration of the target molecule** using a standard curve. ELISAs can generally be categorized into **four**

main groups: **direct**, **indirect**, **sandwich**, and **competitive** ELISAs, each with its specific application and advantages.

Virtual learning aid

Direct ELISA

The simplest ELISA technique involves **immobilizing the antigen from the sample onto the wells of a microtiter plate**. After thorough washing, an enzyme-linked antibody that complements the antigen of interest is added to the wells, forming a bound **antigen-antibody complex** on the well's surface. Subsequently, a substrate is introduced, which is converted by the enzyme-linked antibodies into a detectable product, observable through **color**, **fluorescence**, or **luminescence**. This method is advantageous for its simplicity and efficiency, using just one antibody with fewer steps. However, it has limitations, especially in complex samples with various antigens present, leading to non-specific adsorption and reduced sensitivity. Moreover, the conjugation of antibodies with enzymes might lower the antibody's affinity to the antigen, further impacting sensitivity.

Virtual learning aid

Materials

1. **Antigen sample**(s) - The sample containing the target antigen of interest.
2. **Coating Buffer** - A suitable buffer (e.g., carbonate-bicarbonate buffer) for immobilizing antibodies on the solid surface.
3. **Blocking Buffer** - A blocking agent (e.g., BSA, milk) to prevent non-specific binding.
4. **Primary Antibody** - Specific antibodies against the target antigen.
5. **Secondary Antibody** - Enzyme-conjugated antibodies (e.g., HRP-conjugated anti-mouse IgG).
6. **Substrate** - A chromogenic substrate (e.g., TMB, ABTS) for enzyme detection.
7. **Stop Solution** - An acidic solution to halt the enzyme reaction.
8. **Washing Buffer** - A buffer (e.g., PBS with Tween) for washing steps.

Equipment

1. **Microplate** - 96-well ELISA plate.
2. **Microplate Reader** - Capable of measuring absorbance at specific wavelengths.
3. **Plate Sealer or Adhesive Film** - To cover the microplate during incubation.
4. **Multichannel Pipette** - For efficient dispensing of reagents into multiple wells.
5. **Pipettes and Tips** - For accurate transfer of liquids.
6. **Microplate Washer** - For washing steps.
7. **Incubator** - To maintain a stable temperature during incubation.
8. **Microcentrifuge** - If needed for sample preparation.

Protocol

1. **Preparation of Microplate**: Label the microplate wells accordingly for identification. Coat the wells with the primary antibody (diluted in coating buffer) specific to the target antigen by adding an appropriate volume to each well. Seal the microplate with a plate sealer or adhesive film and incubate at **4°C overnight or 37°C for 2 hours**.

2. **Blocking Step**: Discard the coating solution and **wash** the microplate **3 times** with washing buffer to remove unbound primary antibodies. Add blocking buffer to each well to prevent non-specific binding sites and incubate at **room temperature for 1-2 hours**.

3. **Sample and Standard Addition**: Prepare a series of standard antigen solutions with known concentrations. Add the antigen samples and standard solutions to separate wells in duplicate or triplicate. Include blank wells with only the blocking buffer as a control.

4. **Antibody Binding**: Remove the blocking buffer and **wash** the plate **3 times** with washing buffer. Add the antigen-specific primary antibody (diluted in blocking buffer) to each well and incubate at **room temperature for 1-2 hours**.

5. **Detection Step**: Wash the plate **3 times** with washing buffer to remove unbound primary antibodies. Add the enzyme-labeled secondary antibody (e.g., **HRP-conjugated anti-mouse IgG**) diluted in blocking buffer to each well and incubate at **room temperature for 1 hour**.

6. **Color Development**: **Wash** the plate **3 times** with washing buffer to remove unbound secondary antibodies. Add the **substrate solution** (e.g., **TMB or ABTS**) to each well and incubate for the appropriate time until color development is observed.

7. **Stop Reaction**: Add the **stop solution** (e.g., **2M H2SO4**) to halt the enzyme reaction and stabilize the color.

8. **Measurement**: Measure the absorbance of each well at the appropriate wavelength using a **microplate reader**.

Indirect ELISA

In the **indirect ELISA** method, an enzyme is not directly attached to the antibody binding to the antigen. Instead, the **antibody acts as a bridge between the antigen and an enzyme-linked secondary antibody**. This method is employed **to detect the presence of antibodies against specific viruses**, such as HIV. In this test, antigens are immobilized at the bottom of a well, and patient-derived antibodies (primary antibodies) are added to bind to the antigens. Subsequently, enzyme-linked secondary antibodies specific to the human antibodies are allowed to react in the well, followed by washing to remove any unbound antibodies. The presence of an enzyme reaction indicates that the enzyme-linked antibodies have successfully bound to human antibodies, thus implying the patient's exposure to the viral antigen.

Virtual learning aid

Materials

1. Microtiter plates (96-well format)
2. Coating buffer (e.g., carbonate-bicarbonate buffer, pH 9.6)
3. Blocking buffer (e.g., 1% BSA in PBS)
4. Primary antibody specific to the target molecule
5. Secondary antibody conjugated with an enzyme (e.g., HRP, alkaline phosphatase)
6. Substrate solution suitable for the enzyme (e.g., TMB, ABTS)
7. Stop solution (e.g., 1M H2SO4 for TMB)
8. Wash buffer (e.g., PBS with 0.05% Tween 20)
9. Sample or standard solutions containing the target molecule

Equipment

1. Microplate reader capable of measuring absorbance at desired wavelengths
2. Pipettes (single and multichannel)
3. Microplate washer or aspiration equipment
4. Incubator with temperature control (37°C)
5. Refrigerator (4°C)
6. Disposable plates or tubes for sample dilutions

Protocol

1. **Coating the Microtiter Plate**: Dilute the primary antibody to the desired concentration in the coating buffer. Add **100 µL** of the diluted primary antibody to each well of the microtiter plate. Incubate the plate at **4°C** or **room**

temperature for 1-2 hours or as recommended for the specific primary antibody.

2. **Blocking**: Discard the contents of the plate and **wash** it **three times** with the wash buffer to remove any unbound primary antibody. Add **200 μL** of the blocking buffer to each well. Incubate the plate at **37°C for 1 hour** to block any uncoated surfaces and minimize non-specific binding.

3. **Adding Samples and Standards**: Prepare a series of dilutions of your sample or standard containing the target molecule in the sample buffer. Add **100 μL** in each dilution to the appropriate wells of the microtiter plate in duplicate. Include blank wells containing the sample buffer only.

4. **Binding of Secondary Antibody**: Dilute the secondary antibody in the blocking buffer as per the manufacturer's recommendations. Add **100 μL** of the diluted secondary antibody to each well. Incubate the plate at **37°C for 1 hour** to allow the secondary antibody to bind to the primary antibody.

5. **Washing**: Discard the contents of the plate and **wash** it **five times** with the wash buffer to remove any unbound secondary antibody.

6. **Substrate Incubation**: Prepare the substrate solution according to the manufacturer's instructions. Add **100 μL** of the substrate solution to each well. Incubate the plate at room temperature in the dark for the appropriate time (typically **10-30 minutes**) to allow the color to develop.

7. **Stopping the Reaction**: Add **50-100 μL** of the stop solution to each well to halt the enzymatic reaction. Gently mix the plate to ensure the color development is stopped uniformly.

8. **Measurement of Absorbance**: Using a microplate reader, measure the absorbance of each well at the appropriate wavelength for the substrate used. Record the absorbance values for each sample and standard.

Sandwich ELISA

In a **sandwich ELISA, an antigen is detected or measured by using two specific antibodies**. The first antibody is immobilized on a microtiter well and acts as a capture antibody, binding to the target antigen with high affinity. Next, a separate enzyme-linked antibody that recognizes a different epitope on the same antigen is used to detect the bound antigen. After adding the antigen-containing sample and allowing it to react with the immobilized antibody, the well is washed to remove any unbound components. Then, the second enzyme-linked antibody is added, and after another round of washing, a substrate for the enzyme is introduced. The resulting colored reaction products are measured to determine the presence and quantity of the specific antigen. This method requires two different antigen-specific antibodies that recognize distinct epitopes on the same antigen, enabling accurate and sensitive detection.

Virtual learning aid

Materials

1. Microplate (96-well)
2. Microplate cover or adhesive film
3. Microplate reader
4. Plate shaker or orbital mixer
5. Micropipettes and tips
6. Graduated cylinders or volumetric pipettes
7. Washing buffer
8. Blocking buffer
9. Antigen-specific capture antibody
10. Detection antibody
11. Substrate solution
12. Stop solution

Protocol

Day 1: Coating the Plate with Capture Antibody
1. Dilute the capture antibody in **coating buffer** (usually **PBS**) to the desired concentration (typically **1-10 µg/mL**).
2. Add **100 µl** of the diluted capture antibody to each well of the microplate.

3. Incubate the plate at **4°C or room temperature (RT) for 4-16 hours**, or as recommended by the antibody manufacturer, to allow the capture antibody to immobilize onto the plate.

Day 2: Blocking and Sample Incubation

1. After the incubation, discard the contents of the plate and tap it gently on **absorbent paper** to remove any residual liquid.
2. **Wash** the plate **three times with PBST** to remove the unbound capture antibody.
3. Add **200 µl** of **blocking buffer (BSA)** to each well to prevent nonspecific binding.
4. Incubate the plate at RT for 1-2 hours to block any remaining uncoated surfaces.

Day 2 (Continued): Sample and Standard Addition

1. Prepare a series of standard solutions with known concentrations of the antigen to be measured.
2. Add **100 µl** of each standard solution and the samples (diluted appropriately in blocking buffer) to separate wells in duplicate.
3. Include blank wells with blocking buffer only to serve as a reference for background absorbance.

Day 2 (Continued): Detection Antibody Incubation

1. **Dilute** the detection antibody in blocking buffer to the recommended concentration (usually **0.1-1 µg/mL**).
2. Add **100 µl** of the diluted detection antibody to each well.
3. Incubate the plate at RT for **1-2 hours**, allowing the detection antibody to bind to the captured antigen.

Day 2 (Continued): Washing and Enzyme Reaction

1. **Wash** the plate **four times** with **PBST** to remove the unbound detection antibody.
2. Add **100 µl** of the appropriate enzyme substrate (e.g., **TMB**) to each well.
3. Incubate the plate at RT in the dark for a suitable time (e.g., **15-30 minutes**) to allow the color reaction to develop.

Day 2 (Continued): Stopping the Enzyme Reaction

1. Add **100 µl** of the stop solution (e.g., **2N sulfuric acid**) to each well to stop the enzyme-substrate reaction.
2. Gently tap the plate to ensure thorough mixing.

Analysis: Measure the absorbance of each well at **450 nm** using a microplate reader. Generate a standard curve using the absorbance values of the standards. Determine the antigen concentration in the samples by comparing their absorbance values to the standard curve.

Competitive ELISA

Competitive ELISA, also known as **inhibition ELISA**, is a complex technique where an inhibitor antigen is utilized. It can be adapted in three formats: direct, indirect, and sandwich ELISA. In this method, the antigen of interest and the inhibitor antigen compete for binding to the primary antibody. The procedure involves incubating the unlabeled primary antibody with the sample containing the antigen of interest, leading to the formation of antigen-antibody complexes. As the antibody is in excess compared to the antigen, there are free antibodies remaining.

Next, the antigen-antibody mixture is added to a plate coated with the inhibitor antigen, which can also bind to the primary antibody. The free primary antibodies in the mixture attach to the inhibitor antigen on the plate, while the antigen-antibody complexes do not and are washed off. Subsequently, an enzyme-labeled secondary antibody is introduced to the plate, binding to the primary antibody already bound to the inhibitor antigen. The enzyme-substrate reaction produces a visible signal for detection. The **final signal** obtained is **inversely correlated with the amount of the antigen of interest** in the sample, means **a stronger signal indicates a lower concentration of the antigen in the sample**. This occurs because primary antibodies bound to the sample antigen are washed off, while free primary antibodies are captured by the inhibitor antigen immobilized on the plate and measured by the enzymatic reaction. Another type of competitive ELISA exists, based on antigen capture, where the plate is coated with an unlabeled antibody instead of an antigen.

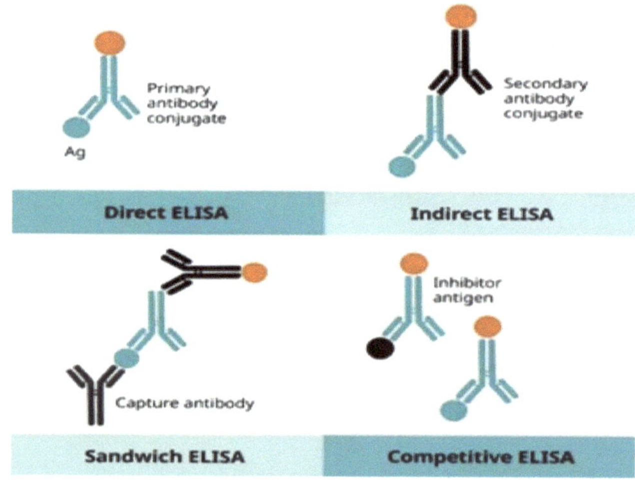

Materials and Equipment

1. **Microtiter Plate**: Flat-bottomed, **96-well polystyrene microtiter plate** (coated with antigen of interest). Well volume: **300-400 μL**

2. **Primary Antibody**: Monoclonal or polyclonal antibody specific to the antigen of interest. Recommended concentration: **1-10 μg/Ml**. Dilution buffer: **PBS** (Phosphate Buffered Saline) or similar

3. **Secondary Antibody**: Enzyme-conjugated secondary antibody (e.g., HRP, alkaline phosphatase) specific to the primary antibody's species (Recommended concentration: **1:500 to 1:2000**). Dilution buffer: **PBS** or similar

4. **Antigen Standard**: Known concentration of the antigen of interest (used for calibration curve). Dilution series: Typically, **5-7 concentrations**, covering the expected range of analyte concentrations

5. **Sample and Control Samples**: Test samples with unknown antigen concentrations (e.g., serum, plasma, cell lysates). **Positive and negative controls**

6. **Blocking Solution**: **Protein-containing buffer** (e.g., BSA, skim milk, or casein). Dilution buffer: **PBS** or similar

7. **Wash Buffer**: **PBS** with **0.05-0.1% Tween 20** or other suitable detergent

8. **Substrate Solution**: **Chromogenic or fluorogenic substrate** for the enzyme conjugated to the secondary antibody

9. **Stop Solution**: Stop enzyme reaction (e.g., stopping HRP reaction using acid or stopping alkaline phosphatase reaction)

10. **Plate Reader**: ELISA plate reader capable of measuring absorbance or fluorescence at appropriate wavelengths

11. **Pipettes and Tips**: Single-channel and multi-channel pipettes with appropriate tips (e.g., 10 μL, 100 μL, 1000 μL)

12. **Microcentrifuge**: For sample preparation and centrifugation steps

13. **Incubator**: To maintain a constant temperature during the assay

14. **Microplate Shaker**: For washing steps

Protocol

1. **Plate Coating**: Dilute the antigen of interest in the coating buffer (e.g., **carbonate-bicarbonate buffer, pH 9.6**). Add the diluted antigen solution to the microtiter plate wells (**100 μL/well**). Cover the plate with adhesive film or a plate sealer and incubate **overnight at 4°C or 2 hours at 37°C** for optimal antigen adsorption.

2. **Blocking**: Discard the coating solution and **wash** the plate **3 times** with wash buffer. Add **blocking solution** (e.g., **1% BSA in PBS**) to each well (**200 μL/well**). Incubate the plate for **1-2 hours at room temperature** to block uncoated sites.

3. **Standards and Samples**: Prepare a dilution series of the antigen standard (e.g., **1:2 serial dilutions**) to generate a calibration curve. Add the antigen standards and test samples (diluted in dilution buffer) to the appropriate wells (**100 μL/well**).

4. **Primary Antibody Incubation**: Dilute the primary antibody in dilution buffer (e.g., PBS) according to the recommended concentration. Add the diluted primary antibody to each well (**100 μL/well**). Incubate the plate for **1-2 hours at room temperature or overnight at 4°C** for antibody-antigen binding.

5. **Washing**: **Wash** the plate **5 times** with wash buffer to remove unbound antibodies and other components.

6. **Secondary Antibody Incubation**: Dilute the enzyme-conjugated secondary antibody in dilution buffer according to the recommended concentration. Add the diluted secondary antibody to each well (**100 μL/well**). Incubate the plate for **1-2 hours at room temperature** to allow the binding of the secondary antibody to the captured primary antibody.

7. **Washing**: **Wash** the plate **5 times** with wash buffer to remove unbound secondary antibodies.

8. **Enzyme Reaction**: Prepare the substrate solution according to the manufacturer's instructions. Add the substrate solution to each well (**100 μL/well**). Incubate the plate in the dark for an appropriate amount of time (e.g., **10-30 minutes**) to allow the enzyme reaction to occur.

9. **Stopping the Reaction**: Add stop solution to each well (**100 μL/well**) to halt the enzyme reaction. Gently mix the plate to ensure complete stopping of the reaction.

10. **Data Collection**: Measure the absorbance or fluorescence of each well using an ELISA plate reader at appropriate wavelengths. Use the standard curve to determine the antigen concentrations in the test samples.

Immunofluorescence

Immunofluorescence is a powerful technique employing **fluorescent-labeled antibodies to detect specific antigens**. Fluorescent molecules absorb light at one wavelength and emit light at another, allowing visualization of immune complexes containing labeled antibodies. This method can be used to visualize antibody-antigen binding in cells or tissue sections. The **two** main types immunofluorescence are **direct**, where the **antibody is directly conjugated with a fluorophore**, and **indirect**, where an unlabeled primary antibody is detected using a **fluorophore-labeled secondary antibody**.

The most commonly used fluorescent dyes in immunofluorescence are **fluorescein** and **rhodamine**, although other dyes like **phycoerythrin** and **phycobiliproteins** are also used. In direct immunofluorescence, the specific antibody is directly linked to a fluorophore, while **indirect immunofluorescence** involves a **two-step process**: an **unlabeled primary antibody** binds to the target, followed by a **fluorophore-labeled secondary antibody** directed against the primary antibody's Fc portion. This versatile technique provides valuable insights into the localization and presence in specific antigens within biological samples.

Immunoelectrophoresis

Immunoelectrophoresis is a powerful immune technique that combines electrophoresis and immuno-precipitation to identify and characterize antigens in complex mixtures. In the first stage, electrophoresis is performed, followed by immunoprecipitation using specific antibodies against the antigens in the second stage, all within the gel itself without removing the antigens. There are various variants of immunoelectrophoresis, including **classical immunoelectrophoresis**, **crossed immunoelectrophoresis**, **rocket immunoelectrophoresis**, and **immunofixation electrophoresis**.

Virtual learning aid

Scan It!

Classical immunoelectrophoresis

Classical immunoelectrophoresis is a laboratory technique that combines electrophoresis and double immunodiffusion **to separate and identify antigens based on their charge**. It is commonly used in clinical laboratories **to detect the presence or absence of antigens in serum samples**. In this method, an antigen mixture is placed in a well at the center of an agar gel and subjected to electrophoresis, causing the components to separate according to their charge. After electrophoresis, antibodies are applied in a trough cut along the side of the gel, allowing the antibodies and separated antigens to diffuse in the agar. When the optimal antigen-to-antibody ratio is achieved, precipitin lines form, facilitating antigen identification.

Materials and Equipment

1. Agarose gel plates
2. Agarose powder
3. Electrophoresis buffer
4. Sample buffer
5. Standard protein markers
6. Serum samples
7. Antiserum
8. Electrophoresis chamber
9. Power supply
10. Staining solution
11. Destaining solution
12. Image documentation system

Protocol

1. **Preparation of Agarose Gel**: Weigh the required amount of agarose powder based on the gel thickness needed (typically **0.8% - 1.5% w/v**). Dissolve the agarose in the electrophoresis buffer by heating and stirring gently. Avoid boiling to prevent bubble formation. Pour the mixture into the gel plate, ensuring no air

bubbles form, and insert a comb to create wells for sample loading. Allow the gel to solidify completely.

2. **Sample Preparation**: Mix serum samples with an appropriate sample buffer in a **1:1 ratio (v/v)**. Prepare duplicate samples for better accuracy. Add standard protein markers to one well to identify protein bands and estimate molecular weights.

3. **Electrophoresis**: Fill the electrophoresis chamber with enough electrophoresis buffer to immerse the gel completely. Carefully load the prepared samples into the wells, ensuring equal volume in each well. Connect the power supply and run the gel at a constant voltage (**~100-150V**) for a suitable duration (e.g., 1-2 hours).

4. **Immunoprecipitation**: After electrophoresis, remove the gel from the plate and wash it with distilled water to remove any residual buffer. Incubate the gel in the appropriate antiserum solution **at 4°C overnight** to allow the formation of antigen-antibody complexes.

5. **Staining and Visualization**: Gently shake off excess antiserum and wash the gel with distilled water. Stain the gel with **Coomassie blue or silver stain** for **about 1 hour**. Destain the gel in a destaining solution until protein bands become visible.

Crossed immunoelectrophoresis

Crossed immunoelectrophoresis, also referred to as **two-dimensional immunoelectrophoresis**, is a two-stage technique used to separate antigens. In the **first stage, antigens are electrophoretically separated in one direction**. Subsequently, in the **second stage, perpendicular electrophoresis is performed in an agarose gel containing specific antibodies to the antigen**. This interaction between antigen and antibody results in the formation of a precipitin arc when they reach equivalence. This technique is particularly valuable for separating proteins with similar electrophoretic mobility.

It is essential not to confuse crossed immunoelectrophoresis with two-dimensional gel electrophoresis. The latter method separates proteins first based on their isoelectric points and then according to their masses. Both techniques serve crucial roles in the realm of immunoelectrophoresis and protein separation, aiding researchers in understanding and analyzing complex antigen-antibody interactions and protein mixtures.

Materials and Equipment

1. Electrophoresis apparatus with power supply

2. Glass plates
3. Agarose or polyacrylamide gel
4. Buffer solution (Tris-glycine or Tris-acetate)
5. Antiserum (specific antibody)
6. Sample containing the protein mixture
7. Staining solution (e.g., Coomassie Brilliant Blue)
8. Destaining solution (e.g., methanol/acetic acid)
9. Glycerol or another mounting medium

Protocol

1. **Preparation of Gel**: Prepare the agarose or polyacrylamide gel according to the desired resolution and thickness. You can find various commercially available pre-cast gels as well. Ensure that the gel is free of any bubbles and evenly poured between glass plates. Allow the gel to solidify.

2. **Preparation of Antiserum**: Obtain the specific antibody (antiserum) against the target protein. Dilute the antiserum to the appropriate working concentration using a buffer solution.

3. **Sample Loading**: Prepare your protein sample by lysing the cells or tissues and extracting the proteins. Mix the protein sample with a suitable buffer to maintain the pH and ionic strength. Load the protein sample onto the gel wells using a fine pipette.

4. **Electrophoresis**: Place the gel in the electrophoresis apparatus, ensuring that the sample wells are aligned with the cathode (negative electrode). Fill the electrophoresis tank with the buffer solution, covering the gel entirely. Apply an electric current (voltage) across the gel according to the manufacturer's recommendations or your experiment's requirements. Allow the proteins to migrate through the gel based on their charge and size. Smaller, more positively charged proteins will migrate faster.

5. **Crosslinking**: After electrophoresis, remove the gel from the apparatus. Incubate the gel in the antiserum (specific antibody) solution for a sufficient amount of time (**1-2 hours**) **at room temperature or 4°C**. This step allows the antigen-antibody complexes to form at the positions where the target proteins are separated in the gel.

6. **Second Electrophoresis (Crossed Electrophoresis)**: Place the crosslinked gel back into the electrophoresis apparatus. Fill the tank with the same buffer solution used in the first electrophoresis. Apply a second electric current,

perpendicular to the original direction, to facilitate the migration of the antigen-antibody complexes.

7. **Staining and Visualization**: After the second electrophoresis, remove the gel from the apparatus. Stain the gel using a suitable staining solution (e.g., Coomassie Brilliant Blue) to visualize the protein-antibody complexes. Destain the gel with a destaining solution (e.g., methanol/acetic acid) to remove the background staining.

8. **Analysis**: Examine the gel for the presence of protein-antibody complexes. You should observe precipitation lines, each representing a specific antigen-antibody complex. Measure the mobility and intensity of the complexes to determine the relative concentration and antigenic properties of the proteins in the sample.

Rocket immunoelectrophoresis

Rocket immunoelectrophoresis, also known as **electroimmunoassay**, is a rapid and reliable method **for quantifying specific protein antigens in a mixture.** In this technique, a **negatively charged antigen is subjected to electrophoresis within a gel containing antibodies**. As the antigen moves through the gel, it interacts with the antibody molecules. Initially, there is an excess of antigen compared to the available antibodies, preventing precipitation. However, as the antigen travels further, it encounters more antibody molecules, leading to the formation of antigen-antibody complexes that precipitate. This creates a distinct rocket-shaped pattern, with the main precipitate at the "head" of the rocket, and finer lines along the "side" formed by a small amount antigen diffusing sideways and quickly reacting with antibodies to precipitate.

Virtual learning aid

Materials and Equipment

1. Electrophoresis apparatus with a power supply
2. Glass plates and spacers
3. Thin agarose gel
4. Serum samples containing antigens
5. Specific antibodies against the target antigen
6. Buffer solutions (Tris-glycine and Tris-barbital)
7. Electrophoresis buffer (Tris-glycine buffer)
8. Staining solution (e.g., Coomassie Brilliant Blue or Silver stain)

Protocol

1. **Preparation of Gel**: Prepare a thin agarose gel by mixing agarose powder with electrophoresis buffer (Tris-glycine buffer). Heat the mixture until the agarose dissolves completely. Pour the agarose solution into the electrophoresis apparatus between the glass plates and spacers. Allow it to solidify.

2. **Sample Loading**: Using a clean applicator, carefully load the serum samples containing antigens into wells made on the gel. Leave a few wells empty for reference and control samples. Ensure that each sample is loaded at a consistent volume to maintain uniformity.

3. **Electrophoresis**: Submerge the gel in the electrophoresis buffer (Tris-glycine buffer) within the electrophoresis apparatus. Connect the apparatus to the power supply and apply a constant electric field (voltage) across the gel. Allow the antigens to migrate through the gel until they reach their respective positions based on their charge and size.

4. **Preparation of Antibody Well**: After the electrophoresis is complete, create shallow wells in the gel perpendicular to the direction of migration, at a suitable distance from the antigen wells. Carefully apply specific antibodies against the target antigens in these wells. Ensure that each well contains a different antibody.

5. **Immunoprecipitation**: Incubate the gel to allow antigen-antibody reactions to occur. This will result in the formation of rocket-shaped immunoprecipitin arcs.

6. **Staining**: Stain the gel with an appropriate staining solution, such as Coomassie Brilliant Blue or Silver stain, to visualize the immunoprecipitin arcs.

7. **Quantification**: Measure the heights of the rocket-shaped immunoprecipitin arcs using appropriate imaging or densitometry software. Create a calibration

curve using standards of known antigen concentrations. Use the calibration curve to quantify the antigen concentration in the sample based on the heights of the arcs.

Immunofixation electrophoresis

Immunofixation electrophoresis (IFE) is a straightforward technique used in immunology. After electrophoresis, antibodies are placed on a gel surface, and when the antigen-antibody balance is reached, they form significant precipitates with their counterparts in the gel. By washing away unbound antibodies and other proteins, only the immunoprecipitates remain in the gel, as they are too large and insoluble to be washed out from the pores. These immunoprecipitates can be directly stained or identified using various methods, such as with **fluorescein, enzymes,** or **radiolabeled primary or secondary antibodies**, aiding in the detection and analysis of specific antigens.

Monoclonal antibodies and Hybridoma technology

When our bodies encounter an infection or are vaccinated, they generate a diverse array of antibodies known as **polyclonal antibodies**. These antibodies recognize different parts of the infectious agent, called **epitopes**, leading to the activation of various B-cells. Consequently, the resulting antisera contains a mixture of antibodies produced by different B-cell clones, each specific to different epitopes. In contrast, monoclonal antibodies (mAb) are a unique type of antibody that comes from a single B-cell clone targeting a specific epitope. The production of monoclonal antibodies is achieved through **hybridoma technology, where a normal B-cell that produces the desired antibody is fused with a cancerous B-cell (myeloma)**. The resulting hybridoma cell line exclusively produces monoclonal antibodies of a single specificity.

Hybridomas are created by fusing antibody-producing B-cells from the spleen with myeloma cells, which can continuously divide, leading to the formation of **immortal cell lines**. These hybridomas combine the characteristics of both parent cells, enabling them to grow indefinitely in laboratory cultures and produce antibodies continuously. The process generating monoclonal antibodies begins by immunizing a mouse with the specific antigen of interest. Over several weeks, B-cells specific to the antigen multiply in the mouse and start producing antibodies. Afterward, spleen tissue, rich in B-cells, is extracted from the mouse and fused with myeloma cells. The resulting hybridomas produce monoclonal antibodies, which hold great value in various scientific and medical applications.

Hybridoma selection involves using a selective medium containing **aminopterin, an antibiotic that inhibits the de novo nucleotide synthesis pathway. Myeloma cells,** engineered to **lack the enzyme Hypoxanthine-**

guanine **phosphoribosyltransferase (HGPRT)**, are **sensitive to aminopterin and die in this medium**. However, hybridomas, formed by fusing lymphocytes with HGPRT-negative myeloma cells, survive as they inherit HGPRT from the lymphocyte parent. The selective medium, known as the **HAT medium**, contains **hypoxanthine** and **thymidine**, providing substrates for the **salvage pathway utilized by normal cells**. This results in the death in unfused lymphocytes and the survival of hybridomas, leading to a pure population of hybrid cells.

In the process of hybridoma selection, **aminopterin**, a synthetic derivative of pterin, acts as **a competitive inhibitor of the enzyme dihydrofolate reductase, crucial for nucleotide synthesis**. Normal cells can use the salvage pathway for nucleotide synthesis and survive in the presence of aminopterin. However, myeloma cells, deficient in HGPRT, cannot utilize the salvage pathway and perish in the aminopterin-containing medium. The fusion in lymphocytes with HGPRT-negative myeloma cells creates **hybridomas**, which **can thrive in the selective medium**, leading to the isolation of pure hybridoma populations suitable for various applications, such as antibody production.

Virtual learning aid

Materials and Equipment

1. Mouse (or any other appropriate animal) for antibody production.
2. Antigen of interest (protein, peptide, or other target molecule).
3. Adjuvant (e.g., Freund's adjuvant) for immunization.
4. Cell culture medium (e.g., RPMI-1640).
5. Fetal bovine serum (FBS) or serum replacement.
6. Hybridoma cells (myeloma cells and B-cells).
7. Hypoxanthine-aminopterin-thymidine (HAT) medium supplement.
8. Polyethylene glycol (PEG) or other cell fusion agents.
9. Selection medium (e.g., HAT-supplemented medium).
10. Enzyme-linked immunosorbent assay (ELISA) kit.
11. Sterile disposable cell culture dishes, flasks, and pipettes.

12. Cell culture incubator (37°C, 5% CO2).
13. Microscope and cell counting equipment.
14. Cryovials and liquid nitrogen storage.

Protocol

1. Immunization

A. Prepare the antigen: Purify or synthesize the antigen of interest to a high level of purity. Verify its identity and functionality through suitable tests like SDS-PAGE, Western blot, or mass spectrometry.

B. Immunization: Inject the antigen into the mouse subcutaneously or intraperitoneally, mixed with the adjuvant (according to the manufacturer's instructions) to enhance the immune response.

C. Boosting: Administer booster doses of the antigen at regular intervals to the mouse to maintain the immune response.

2. Hybridoma Cell Generation

A. Prepare myeloma cells: Culture myeloma cells in a suitable medium and maintain them in the logarithmic growth phase.

B. Harvest splenocytes: After obtaining a satisfactory immune response from the mouse, sacrifice the animal, and harvest the spleen under sterile conditions.

C. Cell fusion: Mix the harvested splenocytes with myeloma cells in the presence of PEG or other cell fusion agents. Incubate and centrifuge the mixture to obtain hybridoma cells.

3. Hybridoma Cell Selection

A. Culture hybridoma cells: Plate the hybridoma cells in cell culture dishes using HAT medium supplement to select hybridomas while inhibiting non-fused myeloma cells and splenocytes.

B. Monitor cell growth: Observe the hybridoma cell growth daily under the microscope. Remove non-hybridoma cells, if any, to maintain a pure culture.

C. Screen for monoclonal antibodies: Test culture supernatants from hybridoma cells using ELISA to identify and select clones producing monoclonal antibodies.

4. Monoclonal Antibody Production

A. Scale-up hybridoma culture: Once monoclonal hybridomas are identified, scale up the culture using suitable culture flasks or bioreactors.

B. Harvest monoclonal antibodies: Collect the culture supernatant and purify the monoclonal antibodies using protein A/G chromatography or other appropriate methods.

Immunoprecipitation

Immunoprecipitation (IP) is a technique used **to isolate specific antigens from cells or tissue lysates by utilizing a particular antibody with affinity for the target.** This method is widely employed to capture proteins and other biomolecules for subsequent analysis through techniques like western blotting. The basic principle is simple: an antibody, often monoclonal, binds to the target antigen in the sample, forming an immune complex. This complex is then captured on a beaded support to separate and purify the antigen of interest. There are **two** primary approaches for capturing the antigen using antigen-specific antibodies: **the direct capture method,** where **pre-immobilized antibodies are used,** and the **indirect capture method,** which involves the **use of free antibodies.**

Direct capture method

In this method, an **antibody** that specifically targets the antigen of interest is **first attached to a solid support,** such as **agarose or magnetic beads.** The beads, now carrying the bound antibodies, are then **mixed with a cell lysate** that contains the desired antigens. The **immune complexes formed are collected** from the lysate, washed to remove any non-specific binding, and then eluted from the solid support. Finally, the collected immune complexes are **analyzed** to study the properties of the target antigen.

Indirect capture method

In this immunoprecipitation method, specific antibodies for a particular protein are directly added to a mixture of proteins in cell lysates. These free antibodies then interact with their respective protein targets. Later, **beads coated with Protein A, Protein G,** or **Protein A/G** are introduced into the mixture, causing the antibodies bound to their targets to attach to the beads. **Protein A/G** is a fusion protein combining IgG binding domains from Protein A and Protein G, each having different affinities for antibodies from different species and subclasses of immunoglobulin Gs. A challenge with using these proteins is that the antibody-antigen complex may be released during elution. To overcome this, **protein cross-linking** using small, **amine-reactive homobifunctional protein cross-linkers** (e.g., DSS or BS3) is employed to covalently attach the antibody to the beads, making it ready for immunoprecipitation.

Immunoprecipitation methods can use either pre-immobilized antibodies on solid beads or free antibodies to form immune complexes. The latter approach is advantageous when the target protein is present in low concentrations, the antibody has a weak binding affinity for the antigen, or the antibody-antigen binding kinetics are slow. **Agarose beads,** which are porous and sponge-like, have been used traditionally but have **largely been replaced by magnetic**

beads, such as **Dyna beads**, for immunoprecipitation. Magnetic beads have a solid and spherical structure, with antibody binding limited to their surfaces. Moreover, magnetic beads are much smaller in size compared to agarose resin, providing a larger surface area for high-capacity antibody binding. Immunoprecipitated complexes can be easily captured with magnetic beads using a micro-centrifuge tube stand fitted with a magnet, allowing the precipitation of the complexes while leaving the supernatant behind.

Co-immunoprecipitation

Variations of immunoprecipitation (IP) techniques are employed **to study the interactions between the primary antigen protein and other molecules like proteins (Co-IP) or nucleic acids (ChIP and RIP)**. Co-immunoprecipitation (Co-IP) is a valuable method to investigate whether two proteins interact under physiological conditions in vitro. It involves using the antibody-precipitated target antigen to co-precipitate its binding partner(s) or associated protein complex from the cellular lysate. Many protein-protein associations within the cell remain intact when the cell is lysed under gentle conditions, allowing the stable partners of the immunoprecipitated protein to also precipitate. **Co-IP** is commonly utilized **to explore the in vivo association between two proteins of interest**. The process involves specific immunoprecipitation of the target protein from cell extracts, followed by fractionation using SDS-PAGE.

Virtual learning aid

Materials and Equipment

1. Protein extraction buffer (e.g., RIPA buffer)
2. Lysis buffer (e.g., NP-40 lysis buffer)
3. Protease inhibitors
4. Phosphatase inhibitors
5. Antibodies against target proteins (primary antibodies)
6. Protein A/G beads or magnetic beads

7. Preclearing control beads (e.g., IgG-conjugated beads)
8. Washing buffer (e.g., Tris-buffered saline with Tween-20, TBST)
9. Elution buffer (e.g., SDS-PAGE sample buffer)
10. SDS-PAGE gels and running buffer
11. Transfer buffer
12. PVDF or nitrocellulose membrane
13. Blocking buffer (e.g., 5% BSA or milk in TBST)
14. Secondary antibodies (conjugated to HRP or fluorophores)
15. Chemiluminescent or fluorescent detection system
16. Western blotting detection reagents

Protocol

1. **Cell Culture and Sample Preparation**: Grow and treat cells according to your experimental design. Wash cells with ice-cold PBS and collect them in a centrifuge tube.

2. **Protein Extraction**: Add appropriate protein extraction buffer supplemented with protease and phosphatase inhibitors to the cell pellet. Incubate on ice for **15 minutes** with occasional vortexing. Centrifuge the lysate at high speed to pellet cell debris, and collect the supernatant containing the protein extract.

3. **Preclearing Step**: Add the pre-clearing control beads (IgG-conjugated beads) to the protein extract and incubate for **1 hour at 4°C** with gentle rotation to reduce nonspecific binding. Centrifuge the mixture to remove the pre-clearing beads and collect the supernatant.

4. **Immunoprecipitation**: Add the primary antibodies against the target proteins to the precleared lysate and incubate **overnight at 4°C** with gentle rotation. Add **protein A/G beads or magnetic beads** to the lysate containing antibody-protein complexes and incubate for **2-4 hours at 4°C** with gentle rotation to capture the immune complexes.

5. **Washing**: Wash the beads three times with ice-cold washing buffer to remove nonspecifically bound proteins and antibodies.

6. **Elution**: Add the elution buffer (e.g., SDS-PAGE sample buffer) and heat the sample to **95°C for 5 minutes** to elute the immunoprecipitated proteins from the beads.

7. **SDS-PAGE and Western Blotting**: Separate the eluted proteins by SDS-PAGE. Transfer proteins from the gel to a **PVDF or nitrocellulose membrane**

using a semi-dry or wet transfer system. Block the membrane with a blocking buffer to prevent nonspecific binding. Incubate the membrane with appropriate primary and secondary antibodies for specific protein detection. Visualize the protein bands using a **chemiluminescent or fluorescent detection system**.

Chromatin immunoprecipitation

Chromatin immunoprecipitation (ChIP) assays are valuable techniques used **to identify genomic regions that interact with DNA-binding proteins** like transcription factors and histones. By cross-linking protein-DNA complexes, purifying the targeted DNA fragments, and analyzing them via PCR, sequencing, or microarray analysis, ChIP allows us **to study epigenetic modifications and genomic sequences associated with specific regulatory proteins** under particular conditions. In ChIP, cells are first fixed with formaldehyde, which temporarily stabilizes the protein-DNA interactions within the cell. After cell lysis, chromatin is released and fragmented, and antibodies specific to the gene regulatory proteins of interest are used to isolate the DNA fragments cross-linked to those proteins. The purified DNA can then be analyzed to map the

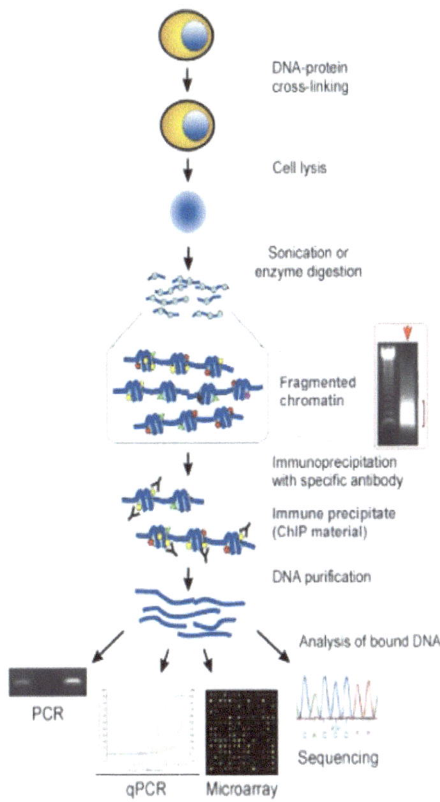

precise genomic locations of the target proteins, aiding in the study of gene regulation and chromatin dynamics in vivo.

Virtual learning aid

Reagents

1. Formaldehyde (37% stock solution)
2. Glycine (2.5 M stock solution)
3. Protease inhibitors
4. Phosphate-buffered saline (PBS)
5. Triton X-100 (1% stock solution)
6. Sodium dodecyl sulfate (SDS) (10% stock solution)
7. Triton/IP buffer (20 mM Tris-HCl, pH 8.0, 150 mM NaCl, 2 mM EDTA, 1% Triton X-100, 0.1% SDS)
8. ChIP dilution buffer (16.7 mM Tris-HCl, pH 8.0, 167 mM NaCl, 1.2 mM EDTA, 1.1% Triton X-100, 0.01% SDS)
9. RIPA buffer (50 mM Tris-HCl, pH 8.0, 150 mM NaCl, 1% NP-40, 0.5% sodium deoxycholate, 0.1% SDS)
10. TE buffer (10 mM Tris-HCl, pH 8.0, 1 mM EDTA)
11. Antibodies:

Equipment

1. Sonicator or Bioruptor (for shearing chromatin)
2. Microcentrifuge tubes
3. Rotator or shaker
4. Microcentrifuge
5. Thermomixer or water bath
6. PCR tubes/strips
7. Real-time PCR machine or qPCR instrument

Protocol

1. **Crosslinking and Cell Lysis**: Harvest the cells of interest and wash them twice with ice-cold PBS. Add formaldehyde to the cell suspension at a final concentration of **1%** and incubate at **room temperature for 10 minutes** with gentle rotation to crosslink proteins with DNA. Quench the crosslinking reaction by adding glycine to a final concentration of **0.125 M**. Incubate for **5 minutes** with gentle rotation. Wash the cells twice with ice-cold PBS, and then resuspend them in Triton/IP buffer supplemented with protease inhibitors. Incubate on ice for **10 minutes** to lyse the cells.

2. **Chromatin Shearing**: Sonicate the cell lysate on ice using a sonicator or Bioruptor to shear the chromatin to an average length of **200-500 base pairs**. Use brief pulses to avoid overheating the sample. Centrifuge the lysate at maximum speed (**at least 12,000 x g**) for **10 minutes at 4°C** to remove cell debris. Transfer the supernatant to a new tube.

3. **Pre-clearing**: Add protein A/G beads to the supernatant and incubate for **1 hour at 4°C** with gentle rotation to pre-clear the lysate from non-specific binding. Pellet the beads by centrifugation and transfer the supernatant to a new tube.

4. **Immunoprecipitation**: Add the specific antibody against the target protein to the pre-cleared chromatin and incubate **overnight at 4°C** with gentle rotation to immunoprecipitate the protein-DNA complexes. Add protein A/G beads to the immunoprecipitated chromatin and incubate for **1-2 hours at 4°C** with gentle rotation to capture the antibody-protein-DNA complexes.

5. **Washing**: Wash the beads sequentially with low salt, high salt, LiCl, and TE buffer to remove non-specifically bound proteins and DNA. Pellet the beads by centrifugation between each wash and remove the supernatant carefully.

6. **Elution**: Elute the protein-DNA complexes from the beads by adding elution buffer (**1% SDS, 0.1 M NaHCO3**) and incubating at **room temperature** with gentle rotation for **30 minutes**. Collect the eluate and reverse the crosslinks by incubating overnight at **65°C**.

7. **DNA Purification**: Treat the eluate with RNase A at **37°C** for **1 hour** to remove RNA. Add proteinase K and incubate at **45°C** for **2 hours** to digest proteins. Purify the ChIP DNA using phenol-chloroform extraction and ethanol precipitation.

RNA immunoprecipitation

RNA immunoprecipitation (RIP) shares similarities with ChIP but **focuses on identifying RNA-binding proteins** instead of DNA-binding proteins. By immunoprecipitating the RNA-protein complexes, followed by **RT-PCR** and **cDNA sequencing**, RIP allows the identification of RNA molecules that interact with specific proteins, shedding light on **post-transcriptional regulatory mechanisms**. Both ChIP and RIP methods provide crucial insights into the dynamic interactions between proteins and DNA/RNA, making them indispensable tools for studying gene regulation, epigenetic modifications, and chromatin organization in various biological contexts. The genomic information obtained through these techniques offers valuable data for understanding gene expression and regulation on a genome-wide scale.

Reagents

1. Protein A/G Magnetic Beads
2. Formaldehyde (37% solution)
3. Glycine (2.5 M, pH 7.4)
4. RNase Inhibitor
5. Protease Inhibitor Cocktail
6. IP Buffer (containing Tris-HCl, NaCl, NP-40, and EDTA)
7. Wash Buffer (containing Tris-HCl, LiCl, NP-40, and EDTA)
8. Elution Buffer (containing Tris-HCl, EDTA, SDS, and proteinase K)
9. Phenol:Chloroform: Isoamyl Alcohol (25:24:1)
10. Chloroform
11. Isopropanol
12. Ethanol (75% and 100%)
13. DEPC-treated Water
14. RNase-free DNase I
15. 10x DNase I Reaction Buffer

Equipment

1. Microcentrifuge
2. Magnetic Stand
3. Sonicator
4. Thermocycler
5. RNase-free Microcentrifuge Tubes
6. RNAse-free pipette tips and tubes
7. RNAse-free Petri dishes
8. UV Crosslinker
9. RNAse-free Water Bath

Protocol

1. **Crosslinking and Cell Lysis**: Harvest cells and wash them twice with ice-cold PBS. Crosslink cells by adding formaldehyde (**1% final concentration**) for **10 minutes** at **room temperature (RT)** with gentle rotation. Quench the crosslinking reaction by adding glycine (**0.125 M final concentration**) and incubate for **5 minutes at RT**. Rinse cells with ice-cold PBS and then lyse them in IP buffer supplemented with RNase and protease inhibitors. Incubate on ice for **10 minutes**. Sonicate the lysate on ice to shear the DNA and achieve an average fragment size of **200-500 nucleotides**.

2. **Immunoprecipitation**: Pre-clear the cell lysate by adding Protein A/G Magnetic Beads and incubate for **1 hour** at **4°C** with rotation. Remove the beads by placing the sample on a magnetic stand and transfer the supernatant to a fresh tube. Add the specific antibody against the target RBP to the supernatant and incubate overnight at **4°C** with rotation. Bind the antibody-RNA complexes by adding Protein A/G Magnetic Beads and incubate for **2 hours** at **4°C** with rotation.

3. **Washes**: Wash the beads **three** times with ice-cold IP wash buffer. After the final wash, resuspend the beads in RNase-free water and transfer them to an RNase-free Petri dish.

4. **Elution and RNA Isolation**: Elute the RNA from the antibody-RNA complexes by adding elution buffer containing proteinase K and incubate for **1 hour** at **55°C** with gentle shaking. Phenol:Chloroform: Isoamyl Alcohol extracts the eluted RNA to remove proteins and phenol. Precipitate the RNA by adding isopropanol and incubate at **-20°C** for **1 hour**. Pellet the RNA by centrifugation at **14,000 rpm** for **30 minutes** at **4°C**. Wash the RNA pellet with **75%** ethanol and centrifuge at **7,500 rpm** for **10 minutes** at **4°C**. Air-dry the pellet and resuspend it in DEPC-treated water.

Chapter 10

Molecular Techniques (Part-1)

Introduction

Nucleic acid extraction is a vital process in biochemical and diagnostic procedures, aiming to obtain pure nucleic acid samples suitable for various downstream applications. The choice of extraction method affects the yield, purity, and extraction time of the obtained samples.

DNA extraction

The choice of **DNA extraction method depends on** various factors, such as the **type of DNA** (chromosomal or plasmid), the **source organism** (bacteria, fungi, plant, or animal), and the **starting material** (organ, tissue, cell, etc.). It also relies on the desired results, including the **molecular weight of the DNA**, **purity**, and **extraction time**. The extraction process involves several steps, including cell lysis (breaking open cells), inactivation of cellular nucleases (enzymes that degrade DNA), removal of other biomolecules from the sample, purification, and quantification of the DNA. Different techniques like mechanical disruption (e.g., grinding, ultrasonication) and chemical lysis (e.g., detergent treatment, alkali treatment) can be used for cell lysis, with chemical lysis employing lytic enzymes, chaotropic agents, and detergents as key components.

Common DNA extraction methods

Nucleic acid extraction techniques can be categorized into **two main types**: **solution-based** and **solid-phase methods**. Solution-based methods involve mixing cell extracts with specialized chemical solutions to purify nucleic acids. On the other hand, solid-phase methods utilize solid supports like magnetic beads coated with materials such as silica to capture nucleic acids. These distinct methods yield varying amounts and levels of DNA purity.

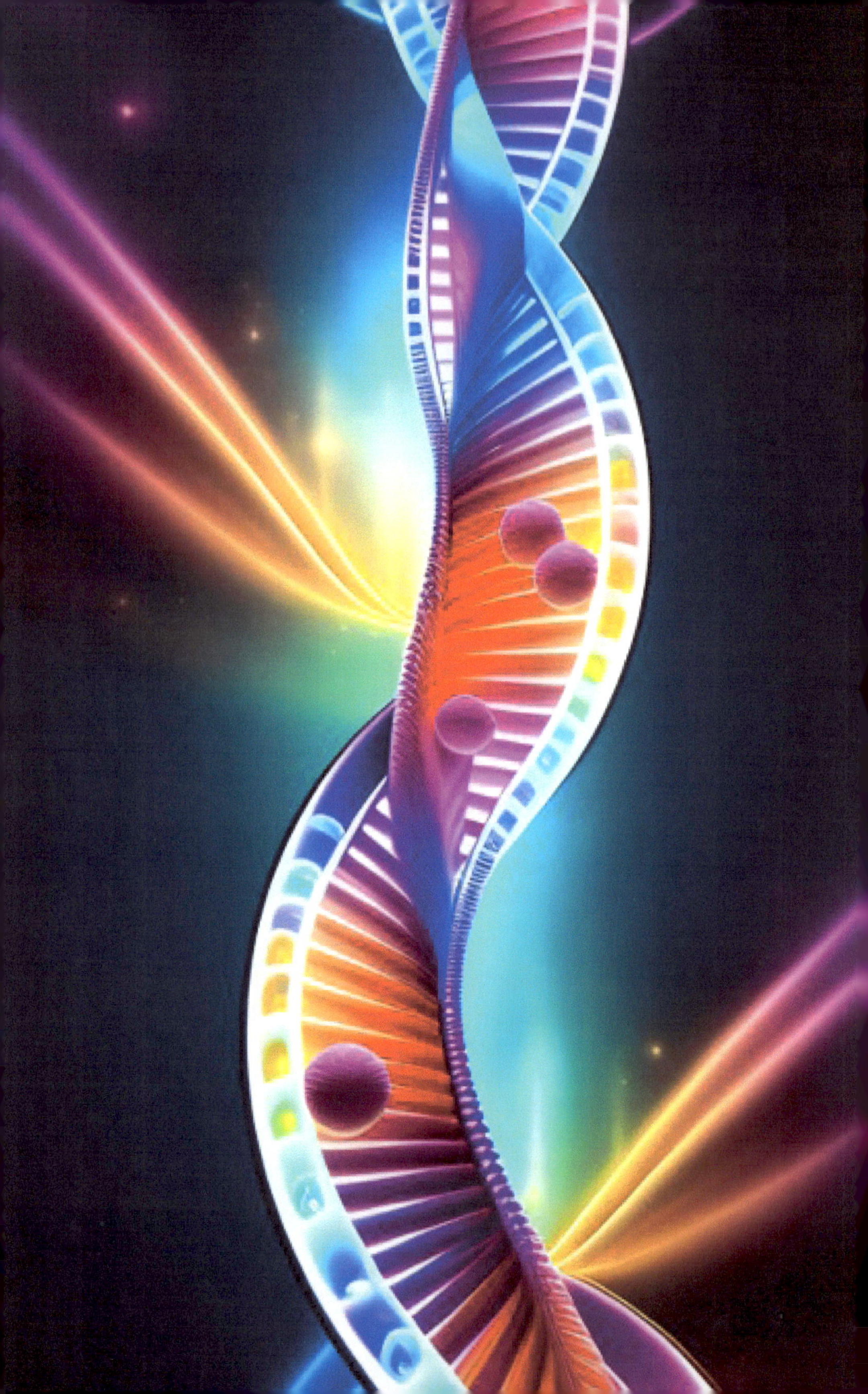

Organic extraction

A traditional method for obtaining pure DNA involves **organic extraction** using substances like **phenol or a mixture of phenol and chloroform**. This process begins with lysing cells and removing debris through centrifugation. The resulting soluble material is combined with organic solvents to concentrate DNA. Phenol dislodges proteins bound to DNA, while chloroform disrupts proteins and lipids. This leads to the formation of **three distinct phases**: an **aqueous phase containing DNA**, an **interphase**, and an **organic phase**. Ethanol is then added to the aqueous phase to precipitate DNA, which can be collected through centrifugation and dissolved in a chosen buffer for downstream applications.

In cases where cell extracts have high protein content, proteases like **pronase** or **proteinase K** are employed **to break down proteins before phenol extraction**. To remove RNA from the preparation, RNase treatment is used. This method efficiently yields purified DNA by separating it from proteins and other contaminants, thus facilitating its use in subsequent experiments.

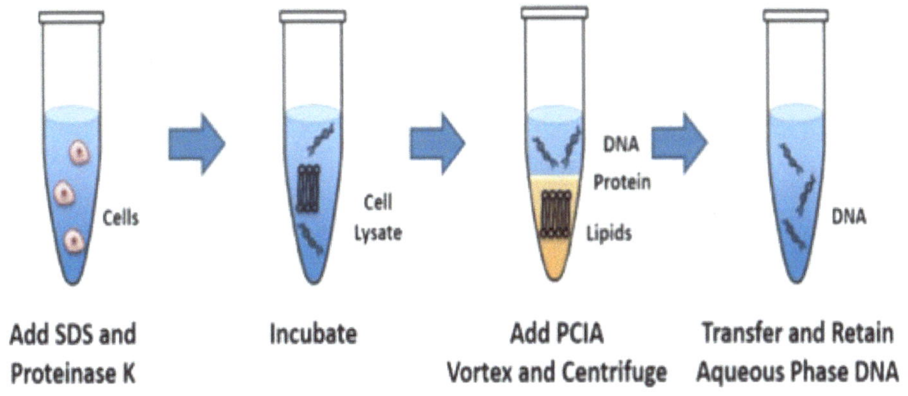

Virtual learning aid

Materials

1. Biological sample (e.g., tissue, blood, cells)
2. Phenol-chloroform-isoamyl alcohol solution (25:24:1, pH 7.9-8.1)
3. Chloroform
4. Isopropanol
5. Ethanol (70% and 100%)
6. TE buffer (10 mM Tris-HCl, 1 mM EDTA, pH 8.0)
7. RNase A (optional, for RNA removal)
8. Proteinase K
9. DNA extraction tubes
10. Microcentrifuge tubes

Equipment

1. Centrifuge
2. Vortex mixer
3. Microcentrifuge
4. Heat block or water bath
5. Pipettes and pipette tips
6. Microcentrifuge tubes
7. UV spectrophotometer (for quantification)

Protocol

1. **Sample Preparation**: Collect the biological sample (e.g., tissue, blood, cells) and transfer it to a DNA extraction tube. Add Proteinase K to the sample according to the manufacturer's instructions (typically **100-200 µg/mL**). Incubate the sample at 55-60°C for 1-2 hours, or until the tissue is fully digested.

2. **DNA Extraction**: Add an equal volume of phenol-chloroform-isoamyl alcohol solution to the digested sample. Close the tube securely and vortex vigorously for **15-30 seconds** to ensure thorough mixing. Centrifuge the tube at **12,000 x g** for **10 minutes** at **room temperature**. Carefully transfer the upper aqueous phase (containing DNA) to a new DNA extraction tube.

3. **Precipitation**: Add an equal volume of chloroform to the aqueous phase, close the tube securely, and vortex. Centrifuge at **12,000 x g** for **10 minutes** at **room temperature**. Transfer the upper aqueous phase to a new DNA extraction tube. Precipitate the DNA by adding **0.6** volume of isopropanol and gently mixing by inverting the tube.

4. **DNA Purification**: Centrifuge at **12,000 x g** for **10 minutes** at **room temperature**. Carefully pour off the supernatant without disturbing the DNA pellet. Wash the DNA pellet with **70%** ethanol by adding the ethanol, inverting the tube, and centrifuging briefly.

5. **DNA Resuspension**: Allow the DNA pellet to air dry for a few minutes. Add TE buffer to the pellet and gently vortex to resuspend the DNA. Incubate the sample at **55-60°C** for **15-30 minutes** to ensure complete dissolution.

Silica-based method

This DNA extraction method employs **solid-phase techniques**. **Silica**, with its **strong attraction to DNA** in alkaline and high-salt conditions, serves as the binding agent. This process relies on the affinity between negatively charged DNA and positively charged silica, which leads to the selective capture of nucleic acids on silica matrices, often in gel or glass particle form. The final step involves eluting DNA from the silica matrix using a **hypoosmotic solution** like nuclease-free water or alkaline Tris-EDTA buffers.

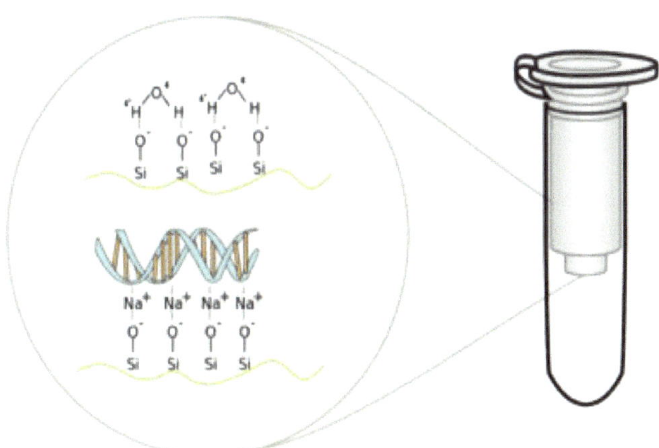

Reagents

1. Phenol-Chloroform-Isoamyl Alcohol (25:24:1)
2. Chloroform
3. Isopropanol (99.8%)
4. Ethanol (70% and 100%)
5. TE Buffer (10mM Tris-HCl, 1mM EDTA, pH 8.0)
6. Silica Binding Buffer: (6M Guanidine-HCl, 10mM Tris-HCl, pH 6.4, 20% Triton X-100)
7. Proteinase K Solution: (20mg/ml in TE buffer)

Equipment

1. Centrifuge
2. Microcentrifuge tubes
3. Pipettes and tips
4. Vortex mixer
5. Heating block or water bath
6. Microcentrifuge with rotor suitable for silica columns
7. UV-Vis spectrophotometer

Protocol

1. **Cell Lysis and Protein Degradation**: Collect the sample (e.g., tissue, blood) and homogenize if necessary. Add **180µl** of Silica Binding Buffer and **20µl** Proteinase K to **200µl** of sample. Mix by vortexing. Incubate at **56°C** for **1-2 hours** for complete cell lysis and protein degradation.

2. **DNA Binding to Silica**: Add an equal volume of Phenol-Chloroform-Isoamyl Alcohol to the lysate. Mix gently and centrifuge. Transfer the aqueous phase to a new tube. Add **0.6x** volume of Isopropanol to the aqueous phase. Mix gently and centrifuge. Discard the supernatant and wash the DNA pellet with **70%** ethanol. Air dry briefly.

3. **DNA Purification**: Resuspend the DNA pellet in **50µl** TE buffer or water. Load the DNA onto a silica column and centrifuge. Wash the column with Wash Buffer (**10mM Tris-HCl, 80% Ethanol**) and centrifuge.

4. **Elution of DNA**: Place the column in a new microcentrifuge tube. Add **30-50µl** of pre-warmed TE buffer (**pH 8.0**) directly onto the column membrane. Incubate for **5 minutes** at **room temperature** and centrifuge.

5. **Quantification and Quality Check**: Measure the DNA concentration using a UV-Vis spectrophotometer. Assess DNA quality through agarose gel electrophoresis.

Magnetic bead-based method

Magnetic beads, possessing uniform particle sizes and **coated with DNA-interacting functional groups**, serve as paramagnetic materials. This **enables DNA to selectively adhere to its surface**, while impurities remain in the solution. By applying an external magnetic field, these beads are separated from contaminants after DNA binding, eliminating the need for centrifugation or vacuum methods. This approach ensures minimal stress on the target DNA molecules.

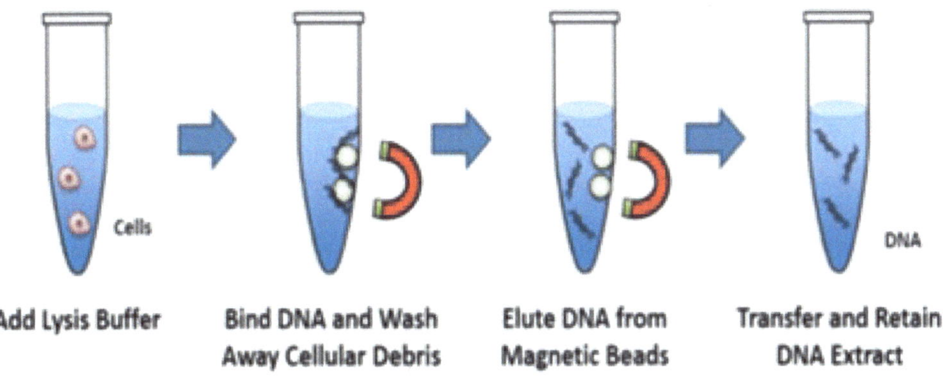

Virtual learning aid

Materials and Equipment

1. DNA sample (e.g., cell lysate, tissue homogenate)
2. Magnetic beads coated with DNA-binding molecules (e.g., silica-coated beads)
3. Lysis buffer (containing detergent and protease)
4. Wash buffer (ethanol-based)
5. Elution buffer (Tris-EDTA or water)
6. Microcentrifuge tubes (1.5 mL and 2 mL)

7. Magnetic stand or rack
8. Magnetic separator
9. Pipettes and pipette tips (varying sizes)
10. Vortex mixer
11. Centrifuge
12. Heating block or thermal cycler
13. Sterile water

Protocol

1. **Sample Preparation**: Collect the biological sample (e.g., cells, tissue) and homogenize or lyse it using an appropriate lysis buffer to release cellular contents.

2. **DNA Binding to Magnetic Beads**: Transfer a desired volume of the lysed sample to a clean microcentrifuge tube. Add an equal volume of binding buffer containing the magnetic beads to the sample. Mix thoroughly by gentle pipetting or vortexing. Incubate the mixture at room temperature for X minutes (refer to manufacturer's recommendations).

3. **Magnetic Bead Separation and Washing**: Place the tube in a magnetic stand to immobilize the magnetic beads. Carefully aspirate and discard the supernatant. Add wash buffer to the bead pellet, resuspend the beads, and incubate for X minutes. Use the magnetic separator to immobilize the beads and remove the wash buffer. Repeat the wash step for a total of Y times.

4. **Elution of DNA**: Remove the tube from the magnetic stand and add elution buffer to the bead pellet. Resuspend the beads in the elution buffer and incubate at Z°C for X minutes. Place the tube in the magnetic stand, and collect the eluted DNA-containing supernatant into a new microcentrifuge tube.

5. **DNA Quantification and Quality Check**: Measure the concentration of the extracted DNA using a spectrophotometer. Optionally, run the DNA on an agarose gel to assess its size and integrity.

Anion exchange method

Anion exchange resins, akin to silica matrices, find extensive application in DNA extraction. This process hinges on the specific interaction between the negatively charged phosphates present in nucleic acids and the positively charged molecules on the substrate's surface. In contrast to the silica's negative charge, anion exchange resin employs **diethyl aminoethyl (DEAE) cellulose** with positive charges to attract nucleic acid's phosphate. In the presence of low salt, DNA selectively binds to the substrate, followed by washing away contaminants using low to medium salt buffers. Ultimately, purified DNA is eluted using a high salt buffer. Notably, anion exchange delivers notably pure DNA in comparison to silica, and the resin can be regenerated for reuse. However, this method employs a high-salt elution step, necessitating desalination for downstream applications.

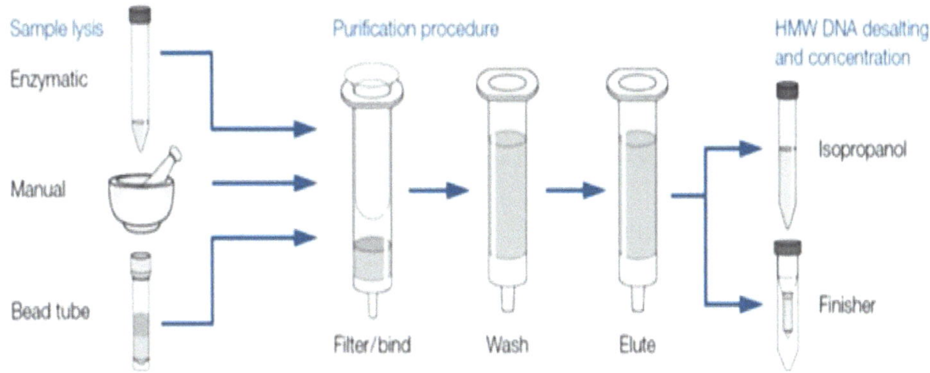

Materials

1. Cell or tissue sample containing DNA
2. Lysis buffer (10 mM Tris-HCl, 400 mM NaCl, 2 mM EDTA, 0.5% SDS)
3. Anion exchange resin (e.g., DEAE-cellulose)
4. Wash buffer I (50 mM Tris-HCl, 50 mM NaCl, 5 mM MgCl2, pH 8.0)
5. Wash buffer II (10 mM Tris-HCl, 1 mM EDTA, pH 8.0)
6. Elution buffer (10 mM Tris-HCl, 1 mM EDTA, 1 M NaCl, pH 8.0)
7. Isopropanol
8. Ethanol (70% and 100%)
9. RNase A (optional)
10. TE buffer (10 mM Tris-HCl, 1 mM EDTA, pH 8.0)

Equipment

1. Centrifuge
2. Microcentrifuge tubes

3. Pipettes and tips
4. Heating block or water bath
5. pH meter
6. Incubator
7. UV spectrophotometer

Protocol

1. **Sample Preparation**: Collect the cell or tissue sample and transfer it to a microcentrifuge tube. Add an appropriate volume of lysis buffer to the sample, ensuring complete coverage. If using tissue, homogenize the sample to ensure thorough mixing.

2. **Cell Lysis and Protein Denaturation**: Incubate the sample at **55-60°C** for **30-60 minutes** in a water bath or heating block to allow cell lysis and protein denaturation. Optionally, add RNase A to the sample and incubate for an additional **15-30 minutes** at **37°C** to digest RNA.

3. **Binding DNA to Anion Exchange Resin**: Prepare a column by packing anion exchange resin (e.g., DEAE-cellulose) into a column according to the manufacturer's instructions. Load the lysed sample onto the column. Wash the column with wash buffer I to remove impurities and proteins.

4. **DNA Elution**: Elute the bound DNA by washing the column with elution buffer. Collect the eluted DNA fractions in separate microcentrifuge tubes. Add isopropanol to the eluted DNA fractions to precipitate DNA. Mix gently and incubate at room temperature for **15-30 minutes**. Centrifuge the tubes at **12,000 × g** for **10 minutes** to pellet the DNA.

5. **DNA Purification**: Discard the supernatant and wash the DNA pellet with 70% ethanol to remove residual salts and contaminants. Air-dry the DNA pellet or centrifuge at low speed to remove excess ethanol. Dissolve the purified DNA pellet in TE buffer.

6. **DNA Quantification and Quality Assessment**: Measure the concentration and purity of the extracted DNA using a UV spectrophotometer. DNA should have an **A260/A280** ratio of around **1.8**.

CTAB-based method

The cetyltrimethylammonium bromide (CTAB) extraction technique is primarily employed for extracting genetic material from plant samples. The basic **CTAB extraction buffer contains 2% CTAB** at an alkaline pH. CTAB works by causing nucleic acids and acidic polysaccharides to precipitate in low-ionic-strength solutions, while proteins and neutral polysaccharides remain in the solution. Subsequently, the CTAB-nucleic acid precipitate complex is dissolved at high salt concentrations, leaving the acidic polysaccharides in the precipitate. Throughout the precipitation and washing steps, the CTAB method utilizes organic solvents like phenol, chloroform, isoamyl alcohol, and mercaptoethanol.

Reagents

1. CTAB (Cetyltrimethylammonium bromide) solution
2. 2-Mercaptoethanol
3. Potassium acetate (KAc) solution
4. Chloroform: isoamyl alcohol (24:1)
5. Isopropanol
6. Ethanol (70% and 100%)
7. TE buffer (10 mM Tris-HCl, 1 mM EDTA, pH 8.0)

Equipment

1. Mortar and pestle (for plant tissue)
2. Microcentrifuge tubes
3. Pipettes (adjustable volume)
4. Vortex mixer
5. Microcentrifuge
6. Water bath or heat block
7. Refrigerated centrifuge

Protocol

1. **Sample Preparation**: Grind or homogenize the tissue sample (e.g., plant material) in liquid nitrogen using a mortar and pestle. Transfer the ground sample into a microcentrifuge tube.

2. **Cell Lysis**: Add **500 μL** of pre-warmed CTAB buffer to the microcentrifuge tube containing the ground sample. Add **10 μL** of 2-mercaptoethanol to the tube and mix by inverting the tube several times. Incubate the tube in a water bath or heat block at **65°C for 60 minutes**, mixing gently every **10-15 minutes**.

3. **Protein Precipitation and DNA Extraction**: Add an equal volume chloroform: isoamyl alcohol (**24:1**) to the tube and mix thoroughly by inverting. Centrifuge the tube at **12,000 × g** for **10 minutes** at room temperature. The mixture will be separated into three phases: an upper aqueous phase, an interphase, and a lower organic phase. Carefully transfer the upper aqueous

phase to a new microcentrifuge tube without disturbing the interphase or organic phase. Add **0.6** volumes of isopropanol to the aqueous phase, mix gently, and incubate at -**20°C for 30 minutes** to precipitate the DNA. Centrifuge the tube at **12,000 × g** for **15 minutes** at **4°C**. The DNA will form a pellet at the bottom of the tube.

4. **DNA Wash and Precipitation**: Discard the supernatant and add **1 mL** ice-cold **70%** ethanol to the DNA pellet. Invert the tube several times to wash the pellet. Centrifuge the tube at **7,500 × g** for **5 minutes** at **4°C**. Carefully remove the ethanol and air-dry the DNA pellet at room temperature for **10-15 minutes**. Resuspend the DNA pellet in **50-100 µL** of TE buffer.

Plasmid DNA isolation

In **plasmid DNA purification**, the separation of plasmid DNA from bacterial chromosomal DNA is **based on their distinct physical properties**. Plasmid DNA and bacterial chromosome DNA, although both circular, **differ in size and structure**. When isolating cellular components, the larger **chromosomal DNA is fragmented** into linear segments with a relaxed conformation. Conversely, due to its compact size, **plasmid DNA maintains a supercoiled structure**. This supercoiling enables an effective separation strategy, such as **alkaline denaturation**. This process exploits the varying responses on chromosomal and plasmid DNA to **sodium hydroxide**, which, when present in high concentrations, denatures both DNA types. However, while linear chromosomal DNA strands separate, the supercoiled plasmid DNA retains its circular integrity. Neutralizing the alkaline conditions with a **potassium acetate buffer** enables the covalently closed plasmid DNA to reanneal, staying soluble, while denatured chromosomal DNA aggregates into an insoluble matrix. Centrifugation then isolates the pure plasmid. DNA in the supernatant, completing the separation process.

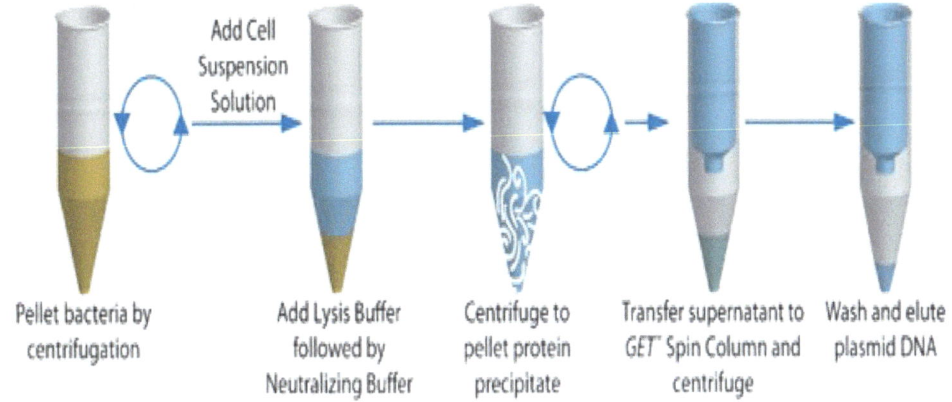

Virtual learning aid

Materials and Equipment

1. Bacterial culture containing the plasmid of interest
2. Lysis buffer
3. RNase A solution
4. Neutralization buffer
5. Binding buffer
6. Wash buffers (ethanol-based)
7. Elution buffer
8. Spin columns
9. Microcentrifuge tubes
10. Centrifuge
11. Pipettes and pipette tips (with sterile filter barriers)
12. Incubator or water bath (37°C)
13. UV-Vis spectrophotometer

Protocol

1. **Preparation of Bacterial Culture**: Inoculate a single colony of bacteria containing the desired plasmid into a suitable growth medium. Incubate the culture overnight at an appropriate temperature.

2. **Harvesting Bacterial Cells**: Transfer the overnight culture to a centrifuge tube. Centrifuge at a predetermined speed and time to pellet the bacterial cells.

3. **Cell Lysis and Plasmid Release**: Resuspend the bacterial pellet in lysis buffer. Mix gently. Add RNase A solution to the lysate to degrade RNA. Incubate at room temperature. Add an appropriate volume of alkaline solution to denature chromosomal DNA. Add neutralization buffer to the lysate to stabilize the plasmid DNA.

4. **Binding and Washing**: Apply the lysate to a spin column preloaded with binding buffer. Centrifuge the column, allowing the plasmid DNA to bind to the column matrix. Wash the column sequentially with wash buffers to remove contaminants.

5. **Elution of Plasmid DNA**: Add elution buffer to the column and centrifuge, eluting the purified plasmid DNA.

6. **Quality Assessment**: Measure the concentration and purity of the isolated plasmid DNA using a UV-Vis spectrophotometer. Optionally, run an agarose gel electrophoresis to visualize the plasmid DNA band.

7. **Storage**: Store purified plasmid DNA at **-20°C for short-term use** or **-80°C for long-term storage**.

Note

One way to separate DNA molecules is by using a technique called buoyant density gradient centrifugation with the help of a DNA-intercalating dye called ethidium bromide (EtBr). When EtBr binds to DNA, it causes a local unwinding of the DNA helix, which decreases the overall density of the DNA. This difference in density affects how much EtBr can bind to different types of DNA. Linear chromosomal DNA, which is not supercoiled, can bind more EtBr compared to supercoiled DNA due to its distinct structural properties. Supercoiled plasmid DNA is resistant to unwinding, leading to less EtBr intercalation. In contrast, linear DNA molecules can accommodate more EtBr, resulting in a reduction of their molecular density.

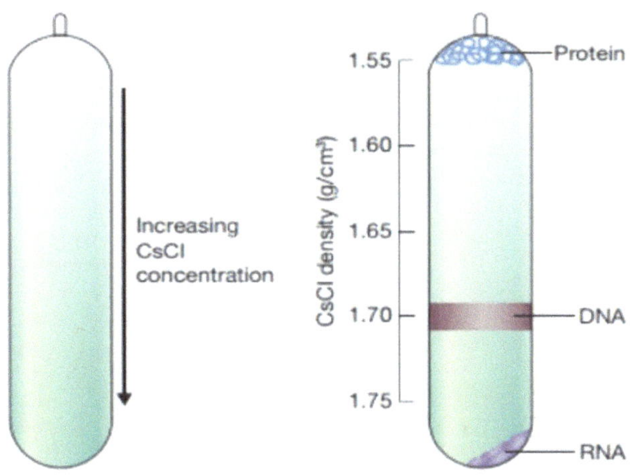

RNA extraction

RNA extraction differs significantly from DNA extraction due to distinct structural and chemical properties. RNA stands as a single-stranded molecule, whereas DNA typically adopts a double-stranded form. This contrast extends to biochemical stability: RNA proves less robust due to the prevalence of RNases, enzymes that degrade RNA, and the inherent instability linked to the ribose ring's 2' hydroxyl group. Even with rigorous RNase removal, RNA can degrade spontaneously in alkaline solutions, necessitating storage at -20°C.

Effective RNA isolation necessitates specific steps: cell or tissue disruption, neutralization of RNase activity, and elimination of DNA and protein impurities. Immediate inactivation of released RNases from organelles upon cell disruption is vital. Among **three primary techniques**, **organic extraction** (such as phenol-guanidine isothiocyanate-based solutions), **silica-membrane spin-columns**, and **paramagnetic particles**, the latter two are favored. They eliminate toxic solvents, offer simplicity and efficiency, and yield pure, intact RNA with minimal contamination.

A widely used method is **phenol-guanidine isothiocyanate-based extraction**, which leverages **Guanidinium isothiocyanate's (GITC)** denaturing power and centrifugation. During centrifugation, the sample segregates into an **organic phase**, a **protein-DNA phase**, and an **aqueous RNA phase**. The upper aqueous layer containing RNA is retrieved and further processed through alcohol precipitation and rehydration, ensuring successful RNA extraction.

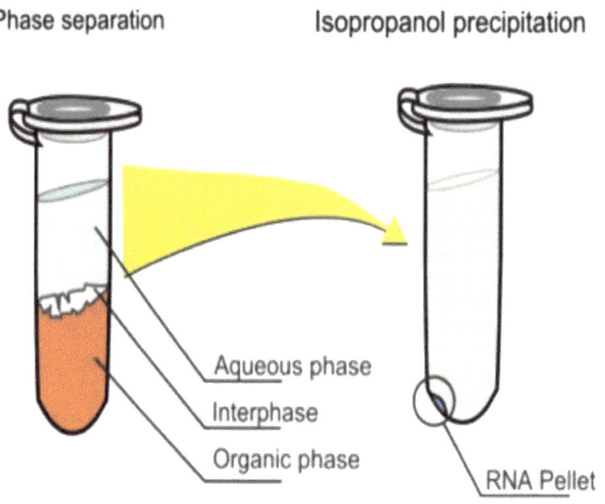

Virtual learning aid

Materials and Equipment

1. Guanidine isothiocyanate solution
2. Phenol-chloroform solution
3. Chloroform

4. Isopropanol
5. Ethanol
6. Centrifuge
7. Microcentrifuge tubes
8. RNAase-free pipette tips and tubes
9. RNase-free water
10. RNase inhibitors

Protocol

1. **Homogenize or lyse** the sample in guanidine isothiocyanate solution, maintaining a 10:1 volume ratio.

2. **Centrifuge** at a specific speed and duration to separate phases: aqueous, interphase, and organic.

3. **Transfer the aqueous phase** to a fresh tube, avoiding the interphase and organic phase.

4. **Precipitate RNA** by adding isopropanol, centrifuging, and discarding the supernatant.

5. **Wash the RNA pellet** with ethanol, then air-dry or centrifuge briefly.

6. **Dissolve the RNA pellet** in RNase-free water. Use RNase inhibitors to prevent degradation.

Trizol Method

The Trizol method, based on the original phenol-guanidine isothiocyanate method, simplifies RNA extraction by combining the reagents into a single solution. Trizol efficiently disrupts cells, denatures proteins, and inactivates RNases, making it suitable for a wide range of sample types.

Materials and Equipment

1. Trizol reagent
2. Chloroform
3. Isopropanol
4. Ethanol
5. Centrifuge
6. Microcentrifuge tubes
7. RNAase-free pipette tips and tubes

8. RNase-free water
9. RNase inhibitors

Protocol

1. **Homogenize or lyse** the sample in Trizol reagent, maintaining a 10:1 volume ratio.

2. **Add chloroform**, mix, and **centrifuge** to separate phases: aqueous, interphase, and organic.

3. Carefully **transfer the clear aqueous phase** to a new tube.

4. **Precipitate RNA** with isopropanol, centrifuge, and discard the supernatant.

5. **Wash the RNA pellet** with ethanol, then air-dry or centrifuge briefly.

6. **Dissolve the RNA pellet** in RNase-free water. Use RNase inhibitors to prevent degradation.

Magnetic Bead-based Method

The Magnetic Bead-based method offers a rapid and automated RNA extraction approach. Magnetic beads are functionalized to selectively bind RNA molecules, allowing for efficient purification with minimal hands-on manipulation.

Materials and Equipment

1. Magnetic bead-based RNA extraction kit
2. Magnet or magnetic stand
3. Centrifuge
4. Microcentrifuge tubes
5. RNAase-free pipette tips and tubes
6. RNase-free water

Protocol

1. **Bind RNA to magnetic beads** through the provided binding solution, maintaining proper binding conditions.

2. Apply the bead-RNA complex to a **magnet or magnetic stand**, capturing the beads against the tube wall.

3. Aspirate and **discard the supernatant**, then wash the beads with a provided wash buffer.

4. **Repeat the wash step**, ensuring the removal of contaminants.
5. **Elute purified RNA** from the beads using an RNase-free elution buffer or water.

6. Use RNase inhibitors to **preserve RNA integrity** if necessary.

Note

Qualitative and quantitative analysis of nucleic acid: Before analyzing nucleic acids, quantifying their quality and quantity is a crucial preliminary step. This is achieved through methods like absorbance, fluorescence, and qPCR. Absorbance-based quantification involves measuring UV light absorption at 260 nm, where nucleotide rings absorb light. It's a quick and simple method with a broad detection range, but it lacks sensitivity compared to fluorescence or qPCR. It can't differentiate between ssDNA, dsDNA, or RNA.

To quantify nucleic acids, their absorption at 260 nm is measured. For dsDNA, 1 unit corresponds to 50 µg DNA, while for ssDNA, it is 33 µg, and for ssRNA, it's 40 µg. Absorbance ratios are used to gauge purity: A260/A280 of ~1.8 for pure DNA and ~2 for pure RNA. Ratios below 1.8 indicate DNA contamination by proteins or organic solvents, while increased ratios reveal RNA contamination.

Fluorescence-based methods use dyes binding to specific nucleic acids. When excited by light, these dyes fluoresce only when bound to their target, indicating the nucleic acid's presence. Such methods are more sensitive than absorbance. A260/A230 values, typically 2.0-2.2, indicate nucleic acid purity. Contaminants absorbing at 230 nm, like salts and detergents, affect this ratio, with fluorescence being a superior detector.

Polymerase Chain Reaction

Polymerase Chain Reaction (PCR) is a rapid and versatile laboratory technique used **to amplify specific DNA sequences**. Developed **by Kary Mullis in 1983**, PCR enables the selective amplification of target DNA sequences from complex DNA sources. This is achieved by designing two short DNA sequences called **primers** that are complementary to the target DNA. These primers guide the synthesis of new DNA strands in a temperature-controlled process.

Key components of PCR include:

1. **Thermostable DNA Polymerase**: An enzyme that catalyzes DNA synthesis.

2. **Primers**: Short DNA sequences that bind to the target DNA and serve as starting points for DNA synthesis.

3. **dNTPs**: Building blocks (nucleotides) required for DNA synthesis (dATP, dCTP, dGTP, dTTP).

4. **Divalent Cations (usually Mg2+)**: Essential cofactor for DNA polymerase activity.

5. **Buffer**: Provides a suitable chemical environment for the DNA polymerase to function effectively.

6. **Template DNA**: The original DNA containing the target sequence to be amplified.

Virtual learning aid

Primer designing

In the process of Polymerase Chain Reaction (PCR), the design of primers holds significant importance as it dictates the targeted amplification. Effective primer design relies on prior knowledge of the target DNA sequence, enabling the creation of specific primers that flank the intended DNA fragment. Ensuring minimal binding to unintended DNA regions is crucial in PCR reactions, necessitating adherence to specific primer design principles. These rules follow:

1. **Primer length**: The length of primers used in PCR is crucial, as overly short primers may bind to unintended sites, leading to undesired products. To exemplify, envision a 10 Mbp DNA molecule subjected to PCR with 8-nucleotide primers. On average, the binding sites for such primers would occur roughly once every 48,656 bp. Consequently, 8-nucleotide primers are unlikely to yield a specific amplification product. In contrast, if the primer length is 15 nucleotides, the expected binding site frequency becomes once every 1,073,741,824 bp. While this exceeds the DNA molecule's length, excessively long primers hinder efficient hybridization to the template DNA due to slower binding rates. **The optimal primer length is generally acknowledged to be 18-20 nucleotides**, striking a balance between specificity and easy binding during annealing.

To calculate primer length, greater specificity is achieved with longer primers. The likelihood of a complementary sequence occurring randomly within a DNA molecule can be determined using the equation:

$$K = [g/2] \times [(1- g)/2]^{(A+T)},$$

Where,

A. **K** represents the **expected frequency of the target sequence** in the DNA molecule,

B. **g** denotes the **relative G+C content** of the DNA molecule,

C. **G, C, A**, and **T** represent the **counts of specific nucleotides in the primer**.

For a double-stranded **genome of size N** (in nucleotides), the anticipated count (n) of primer-complementary sites is given by **n = 2NK**.

2. **Nature of sequences**: In molecular processes like PCR, it's important to **avoid** using sequences with **inverted repeats or self-complementary patterns longer than 3 base pairs**, as they tend to fold into hairpin structures, hindering primer attachment to their intended targets. It's crucial that the two primers used have no overlapping complementarity, as even slight similarity between them, given their high concentration during PCR, can lead to unintended hybrid formation.

3. **Tm values**: The **melting temperature (Tm)** marks the point where half the DNA strands are in a double-stranded form. When using two primers together, their **Tm values should not vary by more than 5°C**, and the Tm of the amplification product should not differ from the primers by more than 10°C.

4. **Base composition**: The **GC content should be between 40 to 60%** with an even distribution of all four nucleotides.

5. **Degenerate primers**: To design PCR primers, sequence information is crucial. **When only partial sequence data** is available or the **complete sequence is unknown**, degenerate primers come into play. These primers consist of a mix of oligonucleotide sequences, with certain positions containing multiple possible bases. This approach generates a range of primers with similar sequences, encompassing all potential nucleotide combinations for a specific protein sequence. **When working with just a protein sequence, degenerate DNA primers are employed**. The protein sequence is reverse-translated to derive the corresponding DNA sequence. Given the genetic code's degeneracy, multiple DNA sequences can correspond to a single polypeptide sequence, primarily in the third codon position. This adaptable sequence is utilized to create degenerate primers.

Reaction cycle

The Polymerase Chain Reaction (PCR) operates as a repetitive cycle, where newly formed DNA strands serve as templates for additional DNA synthesis in the following rounds. This process involves a sequence of three consecutive reactions within each cycle.

1. **Denaturation**: Performs typically at about **93-95°C** for human genomic DNA.

2. **Primer annealing**: The annealing temperature in a PCR reaction should facilitate primer-template binding while preventing mismatched hybrids. Typically, it's set around **5°C below the calculated melting temperature (Tm)** of the primers, ranging from 50°C to 70°C depending on the target's characteristics. One widely used equation, the **Wallace rule**, calculates Tm for perfect 15-20 nucleotide duplexes in high ionic strength solvents like 1M NaCl:

$$Tm = (4\times[G + C]) + (2\times[A + T])$$

Here, **A+T** represents the sum of adenine and thymine nucleotides, while **G+C** refers to guanine and cytosine nucleotides within the primer.

3. **DNA synthesis (or primer extension)**: After primers bind to the DNA template, the next step in PCR involves extending the primers by adding DNA building blocks (dNTPs) using a DNA polymerase enzyme. This enzyme works in the 5' to 3' direction, creating a complementary DNA strand. To optimize this process, the reaction temperature is raised to around **70-75°C** for thermostable DNA polymerases. If the primer annealing temperature is close to the extension temperature, both steps can be combined into a more efficient two-step PCR, eliminating the need to switch between temperatures and speeding up the overall process.

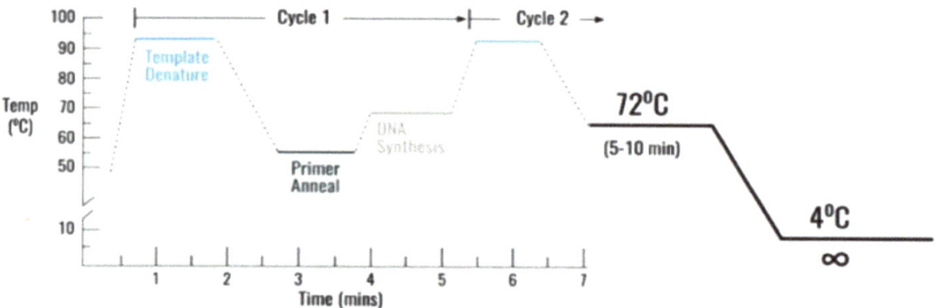

One example is the widely employed **Taq DNA polymerase**, derived from *Thermus aquaticus* bacteria. It remains stable at high temperatures, up to **94°C**, with its best performance occurring between **75-80°C**. Operating at **72°C**, it can elongate DNA strands at a rate of **35 to 100 nucleotides per second.**

However, **Taq lacks a proofreading ability** that corrects errors during DNA synthesis. When it incorporates an incorrect building block, the strand's elongation slows down or halts. To overcome this limitation, alternative thermostable DNA polymerases with 3' to 5' proofreading capabilities can be utilized.

DNA Polymerase	Polymerase Activity	Source Name
DNA Polymerase I	5' to 3' and 3' to 5' exonuclease	*Escherichia coli*
DNA Polymerase III	5' to 3' polymerase	*Escherichia coli*
Taq DNA Polymerase	5' to 3' polymerase	*Thermus aquaticus*
DNA Polymerase alpha	5' to 3' polymerase	**Eukaryotic cells**

After each round of amplification, the number of templates is doubled. For instance, starting with a single dsDNA molecule, after 20 amplification cycles, approximately $1 \times 10\%$ new molecules are generated. After 30 cycles, this number rises to about 1×109 molecules. This relationship can be described using the formula: $N = N_0 \times 2^n$, where **N is the final number of DNA molecules** produced, **N_0 is the initial number of template molecules**, and **n is the number of amplification cycles**.

Nested PCR

Nested PCR enhances the precision of DNA amplification by minimizing unintended DNA replication. It **involves two consecutive PCR rounds**, where the initial round generates DNA products encompassing both the desired target and potential unspecific fragments. In the subsequent round, a different set primers, positioned just downstream or between the first set, is employed. These binding sites diverge partially or completely from the original primers, ensuring focused amplification within the DNA target region.

Materials and Equipment

1. DNA template (genomic DNA or cDNA)
2. DNA primers (outer and nested primers)
3. DNA polymerase enzyme (Taq DNA polymerase)
4. PCR buffer
5. $MgCl_2$ solution
6. dNTP mix

7. Distilled water
8. PCR tubes or plates
9. Thermal cycler (PCR machine)
10. Electrophoresis equipment
11. Agarose gel
12. Gel-loading buffer
13. DNA ladder (size marker)
14. UV transilluminator
15. Gel documentation system

Note

Maintain sterile conditions to prevent contamination.

Protocol

1. **Primer Design**: Design **two sets of primers**: outer primers for the initial amplification and nested primers for the second amplification. Ensure that the nested primers are located within the region amplified by the outer primers to increase specificity.

2. **PCR Setup (Outer PCR)**: Prepare a master mix by calculating the amounts for each component per reaction: **PCR buffer, MgCl$_2$, dNTP mix, forward and reverse outer primers, DNA template, DNA polymerase**, and **distilled water**. Mix the master mix components, except the DNA template and polymerase, in a sterile microcentrifuge tube. Add the DNA template and DNA polymerase to each reaction tube. Place tubes in the thermal cycler.

3. **Outer PCR Amplification**: Perform PCR with the following cycling conditions:
A. **Initial denaturation**: 94°C for 2-5 minutes
B. **Denaturation**: 94°C for 30 seconds
C. **Annealing**: [Annealing temperature]°C for 30 seconds
D. **Extension**: 72°C for 1 minute (adjust the time based on amplicon size)
E. **Repeat steps**: 2-4 for 25-35 cycles
F. **Final extension**: 72°C for 5-10 minutes

4. **Nested PCR Setup**: Prepare a new master mix for the nested PCR with the same components as in Step 2, but using the nested primers. Add the appropriate amount of outer PCR product to the master mix as the DNA template.

5. **Nested PCR Amplification**: Follow the same cycling conditions as in Step 3 for the nested PCR, using the master mix with nested primers and the outer PCR product as the template.

6. **Agarose Gel Electrophoresis**: Prepare a **1-2%** agarose gel in **TAE buffer** and pre-stain it with **ethidium bromide**. Mix PCR products with gel-loading buffer and load them onto the gel, along with a DNA ladder. Run the gel at a suitable voltage until the dye front reaches the desired distance.

7. **Gel Visualization**: Visualize the gel under UV light using a transilluminator. Capture an image of the gel using a gel documentation system.

Quantitative Real-Time PCR

Quantitative Real-Time PCR (qRT-PCR) is a technique rooted in the PCR process, utilized **to both amplify and measure the quantity of a specific DNA segment**. It earns its "**Real-Time**" moniker **due to its ability to visually track DNA amplification** as it occurs. This method employs a **fluorescent reporter**, whose signal intensity corresponds directly to the amount of PCR product generated. Fluorescent reporters like **SYBR Green** bind to double-stranded DNA, emitting minimal fluorescence when unbound but significantly enhancing fluorescence upon binding. While SYBR Green effectively monitors overall double-stranded DNA content, it **exhibits a bias toward G-C rich sequences**. To ensure accurate amplification of the desired target, a sequence-specific fluorescent probe like **TaqMan** is employed. TaqMan probes are short nucleotide strands designed to **bind to internal regions of PCR products**.

The **TaqMan probe** consists of **three parts: a fluorescent molecule, a sequence that matches the target DNA**, and **a quencher**. When the probe is intact, the fluorescent light is blocked by the quencher. The probe binds to the middle of the target DNA. As DNA replication happens during PCR, the probe

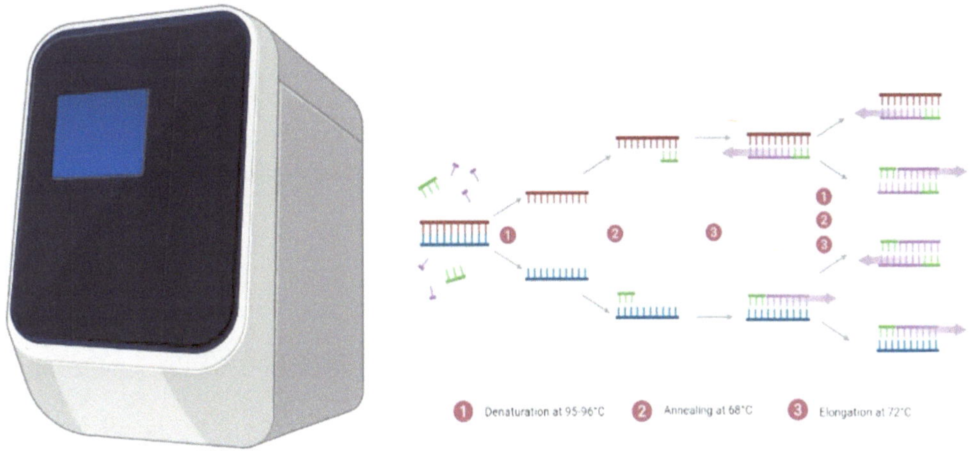

is broken down, separating the fluorescent molecule from the quencher. This allows the fluorescent light to be emitted and measured, indicating the amount of newly formed DNA strands.

Virtual learning aid

Note

Quenchers are substances that can absorb radiation emitted by a fluorophore. They then release a significant portion of this energy either as heat, in the case of dark quenchers, or as visible light, in the case of fluorescence quenchers. This process contributes to the modulation of fluorescence signals and is essential for various scientific applications.

Reagents

1. RNA sample(s)
2. DNase I treatment reagents (if applicable)
3. Reverse transcription reagents (e.g., reverse transcriptase enzyme, random primers, dNTPs)
4. qPCR master mix (including DNA polymerase, buffer, dNTPs, and fluorescent probes)
5. Primer sets for target gene and reference gene
6. nuclease-free water

Equipment

1. Microcentrifuge
2. Thermal cycler (PCR machine)
3. Real-time qPCR machine
4. Pipettes (single and multichannel)
5. RNA extraction kit (if applicable)
6. PCR tubes or plates with optical caps
7. Microcentrifuge tubes

8. Nuclease-free filter tips

Protocol

1. **RNA Extraction and DNase Treatment (if applicable)**: Extract RNA using a suitable method. Check RNA quality and quantity using a spectrophotometer. If required, treat RNA with DNase I to remove genomic DNA contamination.

2. **Reverse Transcription (cDNA Synthesis)**: Prepare a reaction mix containing RNA, reverse transcriptase enzyme, random primers, and dNTPs. Incubate the reaction mix at an appropriate temperature for cDNA synthesis. Inactivate the reverse transcriptase enzyme.

3. **qPCR Setup**: Prepare a qPCR master mix containing DNA polymerase, buffer, dNTPs, and fluorescent probes. Design primer sets for your target gene and a stable reference gene (e.g., housekeeping gene). Add cDNA template, primers, and master mix to PCR tubes or plates.

4. **Real-time qPCR Amplification**: Run the qPCR thermal cycling program, including denaturation, annealing, and extension steps. Monitor fluorescence signals in real-time during each amplification cycle. Collect raw fluorescence data for analysis.

5. **Data Analysis**: Calculate **threshold cycle (Ct)** values for target and reference genes. Determine relative expression using the **ΔΔCt method**, normalizing to the reference gene and a control sample. Perform statistical analysis to assess significance.

RT-PCR

RT-PCR, short for **Reverse Transcription PCR**, is a powerful technique utilized **to amplify**, **isolate**, or **pinpoint specific sequences within cellular or tissue RNA**. Before the PCR step, a reaction involving reverse transcriptase is employed to convert **RNA into complementary DNA (cDNA)**. Remarkably, certain heat-resistant DNA polymerases from *Thermus thermophilus*, like the recombinant **Tth polymerase**, possess the unique ability to **perform high-temperature reverse transcription of RNA templates**. This is particularly notable **when MnCl, a cofactor, is present**. Tth polymerase showcases both intrinsic reverse transcriptase and stable DNA-dependent DNA polymerase activities, enabling the design of protocols that streamline both reverse transcription and PCR amplification within a single enzymatic process.

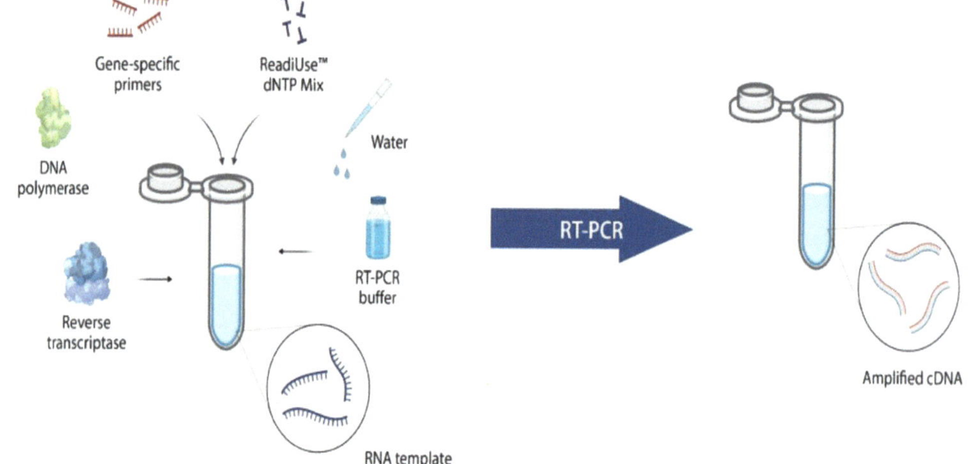

Virtual learning aid

Reagents

1. Total RNA sample
2. DNase/RNase-free water

3. Random primers or gene-specific primers
4. Reverse transcriptase enzyme (e.g., M-MuLV Reverse Transcriptase)
5. Nucleotide mix (dNTPs)
6. RNase inhibitor (optional but recommended)
7. Reaction buffer for reverse transcription
8. PCR primers (forward and reverse)
9. Taq DNA polymerase or a high-fidelity DNA polymerase
10. DNA ladder (molecular weight marker)

Consumables

1. PCR tubes or plates
2. Microcentrifuge tubes
3. Pipettes and pipette tips
4. Nitrile gloves
5. Ethanol and isopropanol for cleaning
6. Adhesive seals or optical adhesive covers

Equipment

1. Thermal cycler (PCR machine)
2. Microcentrifuge
3. Refrigerator and freezer
4. Vortex mixer
5. Water bath or heat block
6. Gel electrophoresis apparatus
7. UV transilluminator
8. Gel documentation system

Protocol

1. **RNA Isolation**: Extract total RNA from your sample using an appropriate method (e.g., TRIzol extraction). Quantify and assess RNA quality using a spectrophotometer or a bioanalyzer.

2. **DNase Treatment (optional)**: Treat RNA with DNase I to remove any contaminating genomic DNA. Inactivate DNase I and purify RNA using standard protocols.

3. **cDNA Synthesis**: Prepare a reaction mix containing RNA, random primers, dNTPs, reverse transcriptase enzyme, and reaction buffer. Incubate the reaction mix at an appropriate temperature (typically **37-42°C**) for cDNA synthesis.

4. **PCR Amplification**: Dilute cDNA and set up PCR reactions with gene-specific primers, dNTPs, DNA polymerase, and reaction buffer. Run the following PCR cycling conditions: **denaturation (94°C, 1-2 min)**, **annealing (varies based on primer Tm)**, and **extension (72°C, varies based on product length)**, for **25-40 cycles**. Include positive and negative controls in your PCR setup.

5. **Gel Electrophoresis**: Prepare an agarose gel with appropriate concentration and stain with a DNA-specific dye. Load PCR products and DNA ladder onto the gel. Run the gel at a suitable voltage until the DNA bands separate.

6. **Visualization and Analysis**: Visualize DNA bands using a UV transilluminator and document using a gel documentation system. Analyze the results, comparing the size of PCR products to the DNA ladder.

Inverse PCR

Standard PCR is a widely used method to amplify a specific segment of DNA between two primers. On the other hand, **inverse PCR**, also known as inverted or inside-out PCR, is employed when we need **to amplify unknown DNA sequences flanking one end of a known DNA sequence without available primers**. In inverse PCR, a series of steps involving restriction digestion and ligation are performed, resulting in circular DNA that can be used for PCR using primers specific to the known sequence. This entails isolating a restriction fragment containing the known sequence and adjacent sequences, which then circularizes through DNA ligase in low concentrations. PCR is carried out with two primers oriented in opposite directions to the known sequence, resulting in a linear product with the central unknown region flanked by two short known sequences.

Reagents

1. DNA template containing the target region.
2. DNA polymerase enzyme (e.g., Taq DNA polymerase).
3. dNTPs (deoxynucleotide triphosphates).
4. Forward primer and Reverse primer (designed based on known sequence).
5. Restriction enzymes (optional, if needed for digestion).
6. DNA ladder (for size comparison).
7. PCR reaction buffer.

Equipment

1. Thermal cycler (PCR machine).

2. Microcentrifuge tubes.
3. DNA electrophoresis setup (gel, buffer, power supply).
4. Pipettes and pipette tips.
5. Water bath or heat block.
6. Incubator (optional, for digestion).
7. Centrifuge (for pelleting DNA after digestion).

Protocol

1. **Design Primers**: Design Forward and Reverse primers flanking the known DNA sequence. Aim for **18-25 base pairs** in length. Ensure the primers have a melting temperature (Tm) of around **50-65°C**.

2. **DNA Extraction**: Extract DNA containing the target region from the sample of interest using a suitable method. Ensure high-quality DNA with minimal contaminants.

3. **Digestion (Optional)**: If necessary, digest the extracted DNA with suitable restriction enzymes to create linear fragments. Use the manufacturer's recommended buffer and conditions. Inactivate the enzymes if required.

4. **Inverse PCR**: Prepare a PCR reaction mix containing a DNA template, primers, DNA polymerase, dNTPs, and PCR buffer. Set up the reaction on ice.

5. **PCR Cycling**: Perform PCR cycling:
A. **Initial denaturation**: 94°C for 2-5 minutes.
B. **Denaturation**: 94°C for 30 seconds.
C. **Annealing**: Adjust temperature based on primer Tm, 30 seconds.
D. **Extension**: 72°C for appropriate time (based on fragment length).
Repeat denaturation, annealing, and extension for a suitable number of cycles (usually 25-35 cycles).
E. **Final extension**: 72°C for 5-10 minutes.

6. **Electrophoresis**: Run the PCR product on an agarose gel with appropriate size marker (DNA ladder). Visualize the amplified fragments using UV light.

Anchored PCR

In the standard PCR and inverse PCR methods, two primers are typically used to amplify a target DNA sequence – these primers represent the sequences at both ends of the target region. However, there are instances where we only know one end of the sequence. In such cases, **anchored PCR** proves useful as it **employs just one primer**. Here, only **one strand of the DNA is initially**

copied, and a poly G tail is added to the new strand's 3'-end. This tailed strand **becomes the template for the next step**, where an anchor primer with a complementary poly C sequence attaches to the poly G tail. Subsequently, both the original primer and the anchor primer are used in the following cycles for gene amplification. This streamlined process allows amplification even when only one end of the target sequence is known.

Reagents

1. Taq DNA Polymerase
2. DNA template
3. Gene-specific primer (GSP)
4. Arbitrary primer (AP)
5. dNTP mix
6. PCR buffer
7. MgCl2
8. Agarose gel
9. Ethidium bromide

Equipment

1. Thermal cycler (PCR machine)
2. Electrophoresis apparatus
3. Power supply
4. Gel documentation system

Protocol

1. **Primer Design**: Design a **gene-specific primer (GSP)** complementary to your target gene sequence. Design an **arbitrary primer (AP)** with a random nucleotide sequence.

2. **Anchoring the Arbitrary Primer**: Mix the DNA template with the AP and heat at **95°C for 5 minutes**, then cool rapidly on ice. This denatures the DNA and allows the AP to anneal at multiple sites. Add dNTP mix, PCR buffer, and Taq DNA Polymerase to the denatured DNA template.

3. **Initial Denaturation**: Place the reaction mix in a thermal cycler. Perform an initial denaturation at 94°C for 2-5 minutes to fully denature the template DNA.

4. **Cycling**: Perform 25-35 cycles of PCR. Each cycle consists of denaturation at 94°C for 30 seconds, annealing at 45-65°C (use the GSP melting temperature) for 30 seconds, and extension at 72°C for 1 minute (adjust times for longer templates).

5. **Final Extension**: After the last cycle, perform a final extension at 72°C for 5-10 minutes to ensure complete extension of PCR products.

6. **Electrophoresis**: Prepare an agarose gel according to standard protocols. Mix PCR products with loading dye and load onto the gel. Run electrophoresis at a suitable voltage until the DNA bands have migrated appropriately.

7. **Visualization**: Stain the gel with ethidium bromide. Visualize the DNA bands using a gel documentation system.

RACE

The **RACE (Rapid Amplification of cDNA Ends)** technique is a PCR-based method utilized to pinpoint the exact beginning and end points of gene transcripts. This method encompasses **two variants**: **5'-RACE** and **3'-RACE**. 1. In **5'-RACE**, mRNA is transformed into cDNA through reverse transcriptase, using a gene-specific primer near the gene's 5'-end. The resulting cDNA precisely corresponds to the mRNA's starting point. Afterward, a short poly(A) tail is added to the cDNA's 3'-end using terminal deoxynucleotidyl transferase, followed by conventional PCR. The second primer binds to this poly(A) tail, initiating the formation of a double-stranded cDNA molecule, which is then amplified during PCR. The sequenced final PCR product reveals the precise start or 5'-end of the transcript, hence the name 5'-RACE due to its focus on amplifying the transcript's 5'-end.

2. In the process known as **3'-RACE**, like 5'-RACE, we need to know the part of the target RNA sequence. First, we convert a group of mRNAs into cDNA using an adapter primer made up of an oligo(dT) primer linked to an adapter sequence. This uses the natural polyA tail found at the 3' end of eukaryotic mRNAs to create a starting point for reverse transcription. After making the first strand of cDNA, we remove the original mRNA template using RNase H, which targets RNA-DNA hybrids. With knowledge of an internal sequence, we use a specific internal primer (gene-specific primer) to build the second DNA strand. To amplify only the 3' end of the cDNA, we employ both the internal primer and a primer matching the adapter sequence in a standard PCR reaction.

Reagents

1. 5' RACE PCR Kit (include brand and catalog number)
2. 3' RACE PCR Kit (include brand and catalog number)
3. RNase Inhibitor
4. RNase-free DNase I

5. RNase-free water
6. dNTP mix (10 mM each)
7. Oligo(dT) primer
8. Gene-specific primers (GSPs) for target gene
9. Universal primer mix
10. Taq DNA polymerase
11. Buffer solutions (1x reaction buffer for RT and PCR)
12. Agarose gel and electrophoresis reagents
13. DNA ladder

Equipment

1. Thermal cycler
2. Centrifuge
3. Gel electrophoresis setup
4. UV transilluminator
5. Pipettes and tips (RNase-free)
6. Microcentrifuge tubes (RNase-free)
7. RNA extraction kit
8. PCR tubes

Protocol

1. **RNA Isolation and cDNA Synthesis**: Isolate total RNA from the target tissue/cells using a reliable RNA extraction kit. Treat the isolated RNA with RNase-free DNase I to remove any genomic DNA contamination. Synthesize cDNA using an oligo(dT) primer and reverse transcriptase (RT). Include an RNase inhibitor in the reaction.

2. **5' RACE**: Design a gene-specific primer (GSP) near the 3' end of the target gene. Perform 5' RACE PCR using the 5' RACE Kit with the cDNA template, GSP, and the provided universal primer mix.

3. **3' RACE**: Design a GSP near the 5' end of the target gene. Perform 3' RACE PCR using the 3' RACE Kit with the cDNA template, GSP, and the provided universal primer mix.

4. **Gel Electrophoresis and Analysis**: Run the PCR products from the 5' and 3' RACE reactions on an agarose gel along with a DNA ladder. Visualize the bands using a UV transilluminator. Excise the desired bands and purify the DNA using a gel extraction kit. Sanger sequence the purified DNA to identify the cDNA ends.

Touchdown PCR

Conventional PCR's accuracy diminishes with DNA templates of higher complexity, increasing the risk of unintended primer binding to other sequences. Adjusting PCR components like Mg2+, primers, dNTPs, and template concentrations can minimize mispriming. A more precise annealing temperature can also reduce off-target amplification. For enhanced specificity and yield, **Touchdown PCR starts with high annealing temperatures, allowing only exact primer-template matches**. Temperature is then gradually lowered, reducing the criticality of high-temperature annealing as earlier products become the main template, which reduces mispriming chances. Lowering the annealing temperature enhances stable primer-target interactions and amplification reliability.

Materials and Equipment

1. DNA template (genomic or plasmid DNA)
2. Forward and reverse primers
3. DNA polymerase (Taq DNA polymerase recommended)
4. dNTP mix (deoxynucleotide triphosphates)
5. PCR reaction buffer
6. MgCl2 solution (if not included in the PCR buffer)
7. Sterile distilled water
8. Thermal cycler
9. PCR tubes/strips and tube caps
10. Microcentrifuge tubes
11. Microcentrifuge
12. Pipettes and tips
13. Electrophoresis setup (optional, for analyzing PCR products)

Protocol

1. **Primer Design**: Design specific primers targeting the DNA region of interest. Ensure suitable primer melting temperatures (Tm) and minimal self-complementarity.

2. **Prepare PCR Reaction Mix**: Calculate the total volume required for the number of reactions. Prepare a master mix by combining the following components for each reaction:
1. **DNA template**: 10-100 ng (adjust as needed)
2. **Forward primer**: 0.5-1 μM
3. **Reverse primer**: 0.5-1 μM
4. **dNTP mix**: 200 μM each

5. **PCR reaction buffer**: 1x concentration
6. **MgCl2** (if needed): optimize concentration
7. **DNA polymerase**: 0.5-1.0 units per reaction
8. **Sterile distilled water** to reach the final reaction volume.

3. **PCR Cycling Conditions**:
A. **Initial denaturation**: 94°C, 2-5 minutes (1 cycle)
B. **Touchdown cycling**:
   a. **Denaturation**: 94°C, 30 seconds (each cycle)
   b. **Annealing**: Start with a higher annealing temperature (T_high), typically 5-10°C above primer Tm. Decrease the annealing temperature by 1-2°C per cycle for 5-10 cycles.
   c. **Extension**: 72°C, 1 minute (adjust time based on product length)
C. **Standard cycling**:
   a. **Denaturation**: 94°C, 30 seconds
   b. **Annealing**: Optimal Tm for primers, 30 seconds
   c. **Extension**: 72°C, 1 minute (adjust time based on product length)
D. **Final extension**: 72°C, 5-10 minutes (1 cycle). Hold at 4-10°C.

4. **Electrophoresis (Optional)**: Analyze PCR products on an agarose gel to confirm successful amplification and estimate product size.

RAPD

Random Amplification of Polymorphic DNA (RAPD), is a variant of PCR that **does not require prior knowledge of a target organism's DNA sequence**. In RAPD, short arbitrary primers (**8-12 nucleotides**) and a large genomic DNA template are used, allowing for comparison of DNA from less-studied biological systems or situations with limited sequence data. This method has been applied to analyze and trace phylogenies in various plant and animal species.

RAPD involves amplifying DNA fragments from different species using a single random primer, without needing prior sequence information. Since RAPD doesn't demand knowledge of the genome being studied, it can be used across species with universal primers. However, RAPD has limitations, including issues with reproducibility and the dominant nature of the markers. Various factors, such as DNA quality, PCR conditions, primer concentration, and annealing temperature, influence the reproducibility of RAPD reactions.

Additionally, RAPD markers are dominant and **do not distinguish between dominant homozygotes and heterozygotes**.

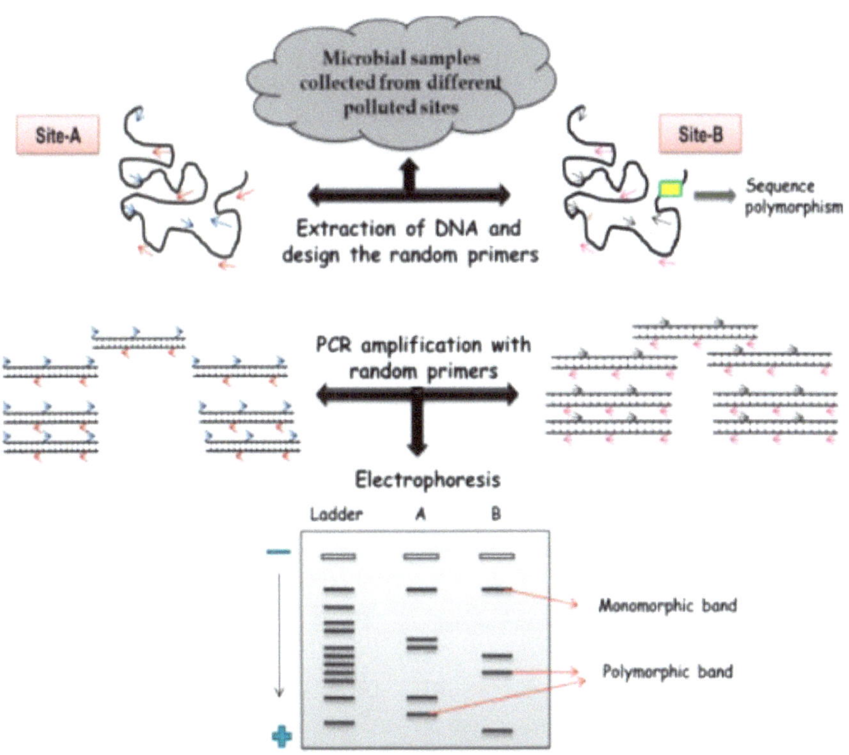

Materials

1. DNA samples (genomic DNA from different sources)
2. DNA primers (random 10-mer primers)
3. Taq DNA polymerase
4. dNTP mix (deoxynucleotide triphosphates)
5. MgCl2 solution
6. PCR buffer
7. Sterile distilled water
8. Loading dye
9. DNA ladder (molecular weight marker)
10. Agarose gel
11. Ethidium bromide solution
12. Gel electrophoresis equipment

Equipment

1. PCR machine (thermal cycler)
2. Centrifuge
3. Water bath
4. Vortex mixer
5. Micropipettes and tips
6. Electrophoresis apparatus
7. UV transilluminator

Protocol

1. **DNA Extraction**: Extract genomic DNA from different sources using a suitable method (e.g., phenol-chloroform extraction or commercial DNA extraction kits). Quantify the extracted DNA using a spectrophotometer or fluorometer.

2. **Primer Selection**: Choose random 10-mer primers that are known to produce polymorphic patterns. Design primer sequences to have a balanced distribution of GC content.

3. **Reaction Setup**: Prepare a PCR reaction mix by combining Taq DNA polymerase, dNTP mix, MgCl2 solution, PCR buffer, and sterile distilled water. Add a DNA template to each reaction tube (e.g., 50 ng genomic DNA). Add a random primer to each reaction (e.g., 1 µM final concentration).

4. **PCR Amplification**: Perform PCR amplification using the following cycling conditions:
A. **Initial denaturation**: 94°C for 5 minutes
B. **Denaturation**: 94°C for 1 minute
C. **Annealing**: 36-40°C for 1 minute
D. **Extension**: 72°C for 2 minutes
Repeat denaturation, annealing, and extension steps for a total of 40 cycles
E. **Final extension**: 72°C for 10 minutes
Use appropriate positive and negative controls.

5. **Electrophoresis and Visualization**: Prepare an agarose gel (1-2%) in TAE or TBE buffer. Mix PCR products with loading dye and load onto the gel, along with a DNA ladder. Perform electrophoresis at a suitable voltage until the dye reaches the desired distance. Stain the gel with ethidium bromide, visualize bands using a UV transilluminator, and capture an image.

AFLP

Amplified Fragment Length Polymorphism (AFLP), technology was developed **to overcome the reproducibility limitations of RAPD**. Unlike RAPD, AFLP **does not require prior DNA sequence knowledge** and is **a dominant marker**. It merges the reliability of RFLP with PCR's adaptability by attaching primer recognition sequences to restricted DNA, followed by selective PCR amplification. Initially, genomic DNA is digested using specific restriction enzymes, and double-stranded adaptors are ligated to ensure unique sequences for PCR. The subsequent PCR amplifications occur only where primers match fragments with adaptor sequences and additional selective nucleotides. This process results in a precise and reproducible analysis of genetic variations.

The **initial PCR step** involves using primers with a **one-base extension**, while the subsequent selective amplification employs primers with **up to a three-base extension**. This high selectivity ensures that **even a single base difference in the AFLP extension leads to the amplification of distinct subsets of fragments**. By extending primers by one, two, or three bases, the number of amplified fragments decreases by factors of 4, 16, and 64 respectively. This optimal adjustment in primer extension yields an ideal number of bands, preventing excessive smearing or a high density of bands. The suitable length for primer extension varies based on the species' genome size and ensures both sufficient polymorphism and avoidance of co-migration during electrophoresis. Visualization of AFLP fragments is achieved through agarose gel or denaturing polyacrylamide gels, often accompanied by autoradiography.

Reagents

1. Restriction enzymes (e.g., EcoRI and MseI)
2. T4 DNA Ligase
3. DNA polymerase
4. dNTP mix (deoxynucleotide triphosphates)
5. Adapter oligonucleotides (EcoRI and MseI adapters)
6. Pre-selective and selective PCR primers (EcoRI and MseI primers)
7. DNA size marker
8. Loading dye

Buffers

1. Restriction enzyme buffer
2. Ligation buffer
3. Pre-selective PCR buffer
4. Selective PCR buffer

Equipment

1. Microcentrifuge
2. Thermal cycler (PCR machine)
3. Electrophoresis apparatus
4. UV transilluminator
5. Gel documentation system

Protocol

1. **DNA Digestion and Ligation**: Extract genomic DNA from samples. Digest the genomic DNA separately with **EcoRI** and **MseI** restriction enzymes in their respective buffers. Inactivate the restriction enzymes and purify the digested DNA fragments. Prepare the EcoRI and MseI adapters by annealing the complementary strands. Ligase the adapters to the digested DNA fragments using **T4 DNA Ligase**. Purify the ligated DNA fragments.

2. **Pre-selective PCR**: Perform a pre-selective PCR using the ligated DNA fragments as templates. Use the pre-selective PCR primers (EcoRI and MseI primers) in this reaction. Purify the pre-amplified products.

3. **Selective PCR**: Perform selective PCR reactions using the pre-amplified products as templates. Use a combination of selective EcoRI and MseI primers with different selective bases to create unique amplification profiles. Prepare a master mix for selective PCR reactions. Amplify the DNA fragments under selective PCR conditions. Purify the amplified fragments.

4. **Electrophoresis and Visualization**: Prepare an agarose gel according to the expected fragment sizes. Load the purified AFLP products and DNA size marker onto the gel. Perform electrophoresis to separate the DNA fragments based on size. Visualize the DNA bands using UV transilluminator. Capture gel images using a gel documentation system.

Nucleic acid hybridization

Nucleic acid hybridization is a fundamental technique in molecular biology that capitalizes on the ability of **single-stranded nucleic acids to pair up and form double-stranded molecules through base pairing**. This process requires a high degree of similarity in the sequence of bases. Hybrids can form between two DNA strands, two RNA strands, or one DNA and one RNA strand. This method is crucial for identifying related molecules based on their complementary base sequences. In hybridization assays, a labeled nucleic acid probe is used to locate matching DNA or RNA sequences within a complex mixture of nucleic acids, known as the target nucleic acid. This technique is not limited to cell extracts; it can also be employed to detect similar DNA or RNA sequences within intact cells or chromosomes, a process known as in situ hybridization.

Stringency of a hybridization reaction

Stringency refers to the **degree of compatibility required between two DNA strands during hybridization**, where they form a double-stranded structure. **Higher stringency** means **fewer mismatches** are allowed before the strands separate. This is influenced by factors like temperature and salt concentration. Stringency **increases** with **higher temperatures** and **lower salt concentrations**, favoring the binding of closely matched sequences. The optimal hybridization temperature is determined experimentally, usually a bit below the temperature at which the DNA strands are 50% separated. Salt concentration also plays a role; low salt promotes separation due to repulsive forces between negatively charged DNA backbones, while high salt stabilizes the double-stranded structure, allowing more mismatches.

The pH of the hybridization solution affects stability too. Slightly acidic conditions enhance stability by reducing repulsive forces between DNA backbones. To achieve successful hybridization, researchers adjust these conditions based on the desired specificity of DNA binding.

Nucleic acid probe

Nucleic acid hybridization using labeled probes is a practical technique to identify specific sequences within complex mixtures of genetic material. These probes, which can be **either DNA or RNA**, are short segments of genetic material that can selectively bind to complementary sequences. Ranging from **20 to hundreds of nucleotides in length**, probes are often labeled for easy detection once they bind to their target nucleic acids. DNA probes can be cloned segments of DNA that are labeled either at their ends or internally during replication, while RNA probes are labeled during transcription from DNA templates. Before hybridization, double-stranded DNA probes are denatured, while RNA and oligonucleotide probes remain single-stranded. Probes can be either similar (**heterologous**) or identical (**homologous**) to the target sequence.

Heterologous probes resemble, but are not identical to, the target sequence of interest. When seeking a gene similar to a known sequence, such as using a mouse probe **to search a human genomic library, heterologous probes are useful**. On the other hand, **homologous probes are exact complements to the target sequence**. The process involves the hybridization of these labeled probes with a mixture of nucleic acids, allowing researchers to identify the presence of specific genetic sequences. This technique is fundamental in molecular biology and genetic research, enabling the precise detection and analysis of genetic material in complex samples.

Note

Labeling of nucleic acids

Nucleic acids can be enhanced with **specialized markers** that facilitate their detection or isolation. These modified nucleic acids are valuable tools for pinpointing or retrieving other molecules that interact with them. Nucleic acids can be labeled by **isotopic** and **non-isotopic labeling** methods:

Isotopic labeling

Isotopic labeling of nucleic acids involves **incorporating radioisotope-containing nucleotides**, a technique known as **radiolabeling**. These labeled probes contain nucleotides with radioisotopes like 32P, 33P, 35S, or 3H, enabling their specific detection in solutions or solid samples. In molecular biology, two key isotopes are vital: radioactive phosphorus-32 (32P) and sulfur-35 (35S). To incorporate 35S into DNA or RNA, phosphorothioate derivatives with radioactive sulfur atoms are utilized, replacing one oxygen in the phosphate group with sulfur. This modification facilitates linking nucleotides together for labeling purposes.

Note

Isotopes are atoms with the same number of protons but different masses. Some isotopes are unstable and emit radiation after nuclear reactions, making them known as radioisotopes or radionuclides.

Isotope	Half-Life
Phosphorus-31	14.28 days
Phosphorus-32	14.262 days
Phosphorus-33	25.34 days
Sulfur-32	89.0 days
Sulfur-33	25.3 seconds
Sulfur-34	4.2 minutes

Radioisotopes emit radiation whose intensity affects their signal strength. This signal can be influenced by exposure time, which can span days or weeks in some cases. For heightened sensitivity, 32p, emitting energetic β-particles, finds use in applications like Southern blot hybridization. Detection of radioisotope incorporation involves two approaches: autoradiography, where radiolabeled material exposes a photographic emulsion, revealing distribution; and direct measurement of radioactivity using Geiger counters (detecting ionization caused by emissions) or scintillation counters (quantifying emitted radiation-induced flashes of light).

In autoradiography, solid sample radioisotope detection on gels or membranes generates an image in a photographic emulsion. Silver halide crystals in gelatin convert emitted β-particles or γ-rays into silver atoms, forming a latent image. **Scintillation counting** relies on scintillants, chemicals absorbing radioisotope emissions and emitting light flashes, detectable by a **photocell**. Radioactive samples mixed with scintillant fluid in a vial are quantified by a scintillation counter, which outputs detected light flashes within a designated time. DNA labeling can occur at ends or uniformly along the molecule.

Labeling DNA by nick translation

Nick translation is a technique used to label DNA segments, involving enzymes like **DNase I**, **DNA polymerase I**, and **DNA ligase**. Initially, DNase I creates breaks at random sites in both strands of the target DNA, followed by DNA polymerase I extending the ends and simultaneously removing bases using its exonuclease activity. This process results in the movement of the nick along the DNA strand. By incorporating radioactive or nonradioactive molecules into the newly synthesized strand, it becomes labeled for various applications. **Nonradioactive methods** involve attaching **digoxigenin** or **biotin** moieties to modified nucleotides.

Random priming

An alternative approach for creating DNA with consistent labeling involves using **hexanucleotides** (or longer sequences) **of random composition to initiate DNA synthesis**. These randomly sequenced oligonucleotides bind to various complementary sites on a DNA fragment, aided by the absence of the 5'-3' exonuclease activity found in DNA polymerase 1. Once bound, they act as starting points for DNA synthesis carried out by DNA polymerases. The Klenow fragment, for instance, fills in gaps between adjacent primers, incorporating labeled nucleotides into the newly synthesized DNA strands.

End-labeling of DNA

End labeling can be performed at the 3'- or 5'-end.

1. **3'-End Labeling**: Calf thymus terminal deoxynucleotidyl transferase catalyzes the addition of [α³²P] NTP to the 3' end of DNA in a template-independent manner. This enzyme, known as TdT, incorporates dNTPs to the 3'-OH end of single or double-stranded DNA and RNA, irreversibly. It's used for creating homopolymeric or heteropolymeric tails at the 3'-end and can also

add a single nucleotide analog like [α^{32}P] cordycepin-5'-triphosphate. Interestingly, TdT is more effective at labeling 3'-protruding ends compared to blunt ends.

2. **5'-End Labeling**: To label the 5'-end, methods involve enzymatic (T4 polynucleotide kinase) or chemical modification of oligonucleotides using phosphoramidite, often using a combination of these approaches. Polynucleotide kinase, usually T4, is employed for 5'-end labeling. This enzyme has two reactions: the forward reaction, where a hydroxyl group is created by removing an unlabeled phosphate from the 5'-end, and the exchange reaction, where the phosphate is transferred to ADP, allowing a labeled gamma phosphate from ATP to be attached to the 5'-end.

Nonradioactive Labeling

Nonisotopic labeling avoids radioactive probes and employs direct and indirect methods. In direct labeling, probes are directly connected to dyes or enzymes that produce the signal. Indirect labeling **uses probes with a hapten that binds to a secondary agent, generating the signal. Two widely used systems** are the **biotin-streptavidin system** and the **digoxigenin system**. Biotin and streptavidin bind strongly, while digoxigenin, a plant steroid, is detected using specific antibodies.

Incorporating Labels

For both biotin and digoxigenin, uracil (normally in RNA) is used in DNA. DeoxyUTP labeled with biotin or digoxigenin replaces thymidine in the DNA chain via DNA polymerase, leaving the labels exposed without affecting the DNA structure.

Molecular beacons

A **molecular beacon** is a specialized fluorescent probe designed with a distinctive **stem-loop structure, offering greater accuracy in target identification compared to linear probes**. This structure involves a loop sequence at the center that perfectly complements the target of interest, while the stem arms are complementary to each other. Comprising around **25 nucleotides**, it contains **a fluorophore** and **a quenching group** at opposite ends. Detection hinges on **Fluorescence Resonance Energy Transfer (FRET)** a phenomenon where energy from an excited fluorophore transfers to an acceptor fluorophore (linked to a quencher dye), resulting in the emission in a fluorescent signal or energy dissipation. Close physical proximity between donor and acceptor molecules is crucial for this energy transfer to occur effectively.

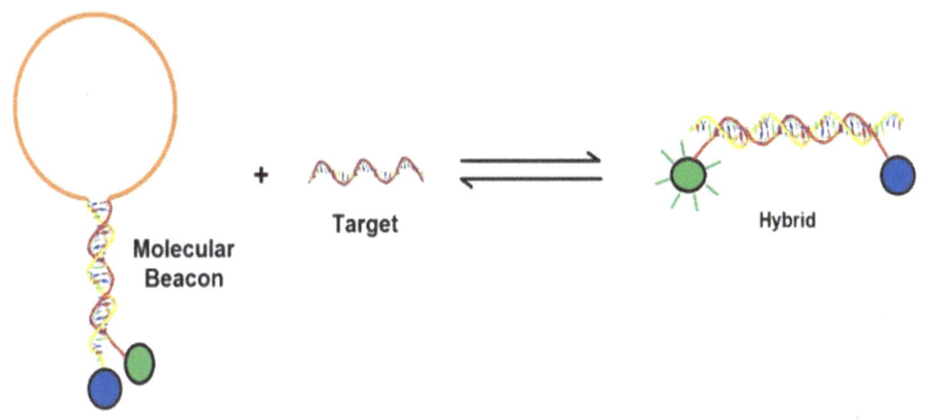

A molecular beacon is a specialized probe that emits fluorescence when it specifically attaches to a certain DNA target sequence. The probe's middle portion complements the target sequence, while the short double-stranded regions at its ends create a stem and loop structure. In this conformation, a quenching group blocks fluorescence alongside the fluorophore. When the beacon binds to the target sequence, it straightens out, separating the quenching group from the fluorophore and allowing fluorescence to occur.

Scorpions

Scorpion probes are special molecules used in the polymerase chain reaction (PCR) process **for detecting and quantifying DNA**. These probes are made up of **two essential parts**: a **PCR primer connected to a probe with a specific structure**. The probe contains a fluorescent dye and a quencher that reduces fluorescence. During PCR, when the DNA is copied, the probe's parts separate, leading to increased fluorescence. There are two main designs for these probes. In one design, the probe forms a stem-loop structure with the fluorescent dye and quencher, while in the other design, the fluorescent probe pairs with another DNA piece carrying the quencher.

In the stem-loop design, the probe includes a primer, quencher, fluorescent dye, and a PCR blocking element. This blocking element prevents the DNA polymerase from extending into the probe during PCR. As the PCR process unfolds, the probe's loop section can bind to a complementary sequence in the target DNA, causing the fluorescent dye to detach from the quencher. This separation produces a measurable fluorescent signal, allowing us to monitor the PCR progress in real time and accurately quantify the DNA produced. This innovative technique aids researchers in various applications requiring DNA analysis.

Fluorescence in situ hybridization (FISH)

Fluorescence In Situ Hybridization (FISH) operates on a principle akin to Southern blot analysis, capitalizing on single-stranded DNA's tendency to pair with complementary DNA sequences. **FISH targets nuclear DNA in interphase or metaphase stages, initially focusing on metaphase chromosomes**. The process entails applying a fluorescent-labeled DNA probe to denatured metaphase chromosome DNA. While radioisotopes were once used for probe labeling, contemporary practices employ fluorescent dyes for heightened sensitivity and resolution. The labeled probe's binding to chromosomal DNA is identified through emitted fluorescent signals. By conducting FISH with two DNA probes, each marked with distinct fluorochromes, the chromosomal markers' positions become visualized, facilitating direct observation of marker locations on chromosomes or extended DNA strands.

The foundation of FISH involves **two primary components**: the **DNA probe** and the **target DNA** for hybridization. The selection of an appropriate probe is a crucial consideration in FISH analysis, ranging from whole genomes to smaller cloned probes. **Probes** fall into **three broad categories**: **whole-chromosome painting probes**, **repetitive sequence probes**, and **locus-specific probes**. Successful hybridization necessitates the denaturation in double-stranded chromosomal DNA to render it single-stranded. This is achieved by applying formamide to the chromosomal preparation, maintaining chromosome structure intact.

FISH visualization predominantly employs **two methods**: **flow cytometry systems** and **slide-based systems**. Flow systems handle FISH-stained cells in suspension, passing them through a laser beam to measure their fluorescent intensities. In slide-based approaches, fixed FISH-stained cells adhere to a microscope slide and are observed statically using wide-field or confocal fluorescence microscopy. Preceding fixation, cells can be arrested in metaphase using colchicine. Subsequent fixation and incubation with solvents and elevated temperatures allow probe DNA hybridization with chromosomes, enabling clear visualization of chromosomal markers through FISH techniques.

Virtual learning aid

Scan It!

Materials

1. Probe DNA: Labeled with a fluorophore (e.g., Cy3, FITC)
2. Target Sample: Fixed cells or tissue sections
3. Hybridization Buffer: Formamide-based buffer (recipe provided below)
4. Blocking Solution: Bovine Serum Albumin (BSA) or Casein solution
5. Wash Buffers: SSC buffers (recipe provided below)
6. DAPI (4',6-diamidino-2-phenylindole): DNA stain
7. Antifade Mounting Medium: To preserve fluorescence signal

Equipment

1. Microscope slides and coverslips
2. Fluorescence microscope
3. Humidity chamber
4. Water bath
5. Oven or incubator
6. Fluorescence filters for desired fluorophores

Hybridization Buffer (Formamide-based)

1. 50% Formamide
2. 10% Dextran Sulfate
3. 2x SSC (Saline-Sodium Citrate) buffer
4. 10 mM Tris-HCl (pH adjusted)
5. 0.1% Sodium Dodecyl Sulfate (SDS)
6. 10% Denhardt's solution
7. 200 µg/mL of sheared salmon sperm DNA
8. 2x SSC buffer

Protocol

1. **Sample Preparation**: Prepare slides with fixed cells or tissue sections adhered. Permeabilize the samples if necessary (e.g., with 0.5% Triton X-100 in PBS). Wash slides in 2x SSC buffer.

2. **Pre-hybridization**: Prepare hybridization buffer by mixing the required components and heating to 75°C to dissolve. Cool the hybridization buffer to 37°C. Apply blocking solution to slides and incubate for 30 minutes at room temperature. Apply probe mixture (probe DNA in hybridization buffer) to each slide. Incubate slides in a humid chamber at 37°C for 2-16 hours.

3. **Post-hybridization Washes**: Prepare wash buffers by diluting 20x SSC to the desired concentration (e.g., 2x SSC, 0.1x SSC) and adding 0.1% SDS. Wash slides sequentially in wash buffers at 37°C to remove unbound probe. Optional: Counterstain nuclei with DAPI (1 µg/mL in wash buffer) for 10 minutes.

4. **Mounting and Imaging**: Apply antifade mounting medium with a coverslip. Seal edges with nail polish. Allow the mounting medium to set.

5. **Imaging and Analysis**: Use appropriate fluorescence filters for your fluorophores. Observe slides under a fluorescence microscope. Capture images and analyze signal patterns and co-localization.

Colony hybridization

In **colony hybridization**, specific DNA-containing bacterial colonies are selected or identified. The process involves transferring bacterial colonies from an agar plate onto a nylon membrane. After releasing DNA from the bacterial cells through alkali hydrolysis and detergent treatment, the DNA binds to the membrane. Single-stranded DNA on the membrane attaches covalently via UV irradiation. This membrane is then exposed to a labeled nucleic acid probe, allowing it to hybridize with its complementary sequence. After thorough washing to remove the unhybridized probe, hybridized regions are visualized. By comparing the membrane with original dish, the original group of colonies can be determined through aligned hybridization regions.

Virtual learning aid

Materials and Equipment

1. Nitrocellulose or nylon membrane filters
2. Sterile filter paper
3. Agar plates with bacterial colonies of interest
4. Alkaline lysing solution (e.g., NaOH)
5. Neutralizing solution (e.g., Tris-HCl)
6. Denaturation solution (e.g., NaOH)
7. Prehybridization buffer
8. Hybridization probe (labeled DNA or RNA)
9. Hybridization buffer
10. Washing buffers (low to high stringency)
11. Autoradiography film or detection system
12. Darkroom or imaging equipment

Protocol

1. **Lysing and Denaturation**: Prepare alkaline lysing solution and neutralizing solution. Transfer the membrane filters to the bacterial colonies on agar plates and gently press to ensure contact. Add alkaline lysing solution on the membrane and incubate briefly. Neutralize with a neutralizing solution and rinse with water. Treat the membrane with a denaturation solution (e.g., NaOH) to denature DNA.

2. **Prehybridization**: Prepare a prehybridization buffer and preheat it. Place the treated membrane in a sealable hybridization bag or container. Add preheated prehybridization buffer and seal the container. Incubate at an appropriate temperature to block non-specific binding sites.

3. **Hybridization**: Prepare the labeled hybridization probe (DNA or RNA) specific to your target sequence. Denature the probe and add it to the hybridization buffer. Remove the prehybridization buffer from the container and add the hybridization buffer with the labeled probe. Seal the container and incubate at a suitable temperature to allow hybridization.

4. **Washing**: Prepare a series of washing buffers with different stringencies (low to high). Perform a series of washes to remove unbound probes and reduce background noise.

5. **Detection and Imaging**: Prepare autoradiography film or set up a suitable detection system. Transfer the membrane to the detection apparatus and expose it to the film or use the detection system. Develop the autoradiography film or analyze the signal obtained from the detection system.

Chapter 11

Sequencing Techniques

Introduction

DNA sequencing refers to the process of determining the exact order of the building blocks, or nucleotide bases (adenine, guanine, cytosine, and thymine), within a DNA molecule. There are several methods for DNA sequencing:

Chain termination method (Dideoxy method)

Sanger sequencing, also known as **dideoxy sequencing**, is a widely used method **for determining the nucleotide sequence of DNA**. This technique relies on the termination of DNA synthesis using modified nucleotides, allowing the identification of the sequence through gel electrophoresis. This method, developed **by Frederick Sanger and colleagues in 1977**, involves the enzymatic synthesis of a complementary DNA strand. Natural deoxynucleotides (dNTPs) and special **dideoxynucleotides (ddNTPs)** that act as terminators are used. DNA synthesis is halted randomly whenever a ddNTP is incorporated into the growing DNA chain, resulting in fragments of varying lengths with a ddNTP at their end. These fragments are separated by size using gel electrophoresis, and the sequence is determined based on the terminal ddNTPs.

Virtual learning aid

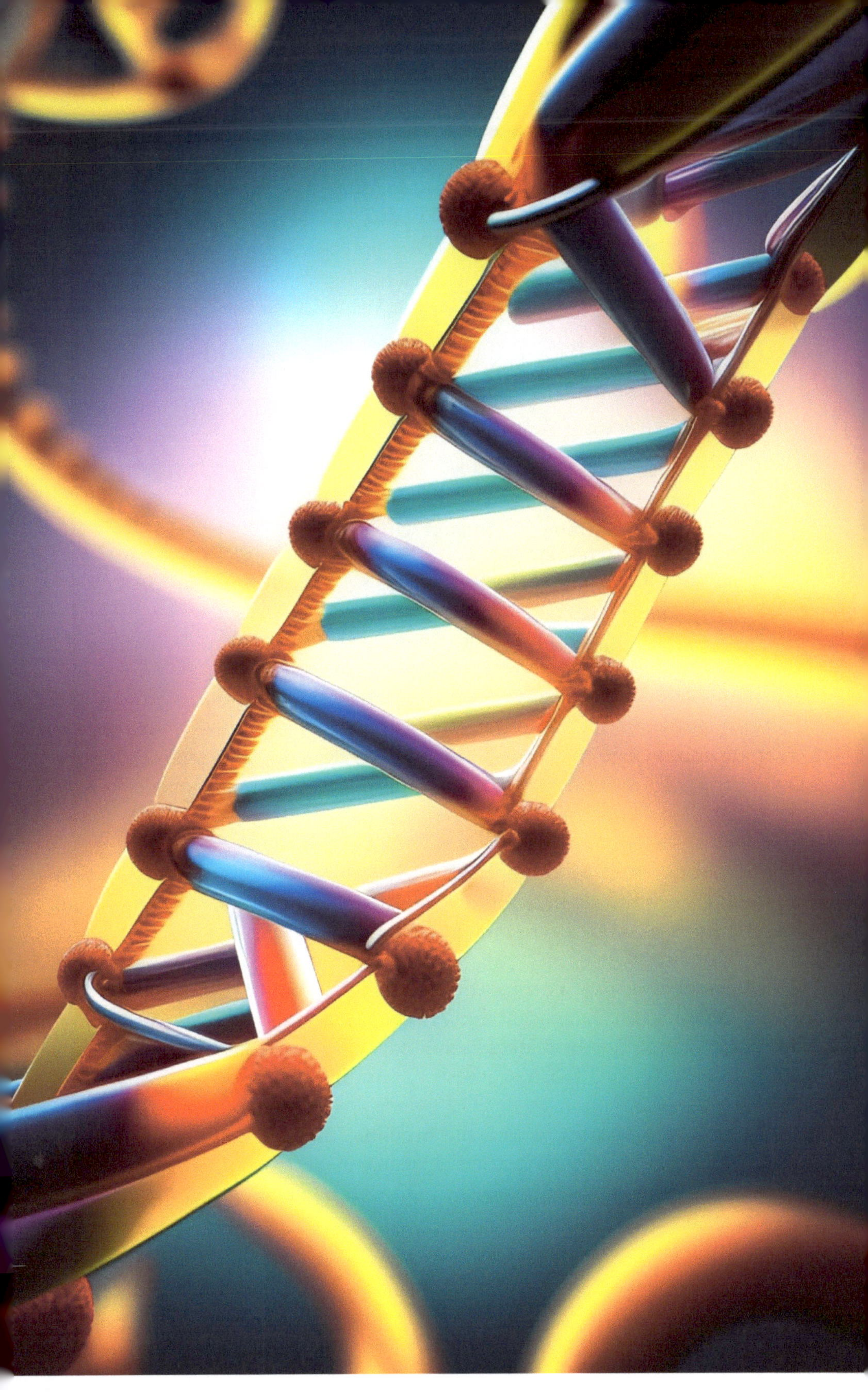

Materials

1. Template DNA (target DNA for sequencing)
2. DNA primer (complementary to the template)
3. DNA polymerase enzyme
4. Four dideoxynucleotide triphosphates (ddNTPs) labeled with distinct fluorescent dyes (ddATP, ddCTP, ddGTP, ddTTP)
5. Deoxynucleotide triphosphates (dNTPs)
6. Reaction buffer
7. Magnesium ions (Mg2+)
8. Gel loading dye
9. Ethidium bromide or other DNA stain
10. DNA ladder or size marker

Equipment

1. Thermal cycler (PCR machine)
2. DNA sequencing machine (automated sequencer)
3. Electrophoresis apparatus
4. Capillary array (for automated sequencing)
5. Fluorescence detector
6. Computer with sequencing software

Protocol

1. **Preparation**: Isolate the template DNA from the sample of interest. Design a DNA primer complementary to the template at the desired sequencing start site.

2. **Primer Annealing**: Mix template DNA, primer, and appropriate reaction components. Heat the mixture to denature the DNA strands Cool the mixture to allow the primer to anneal to the template DNA.

3. **DNA Synthesis and Termination**: Add DNA polymerase, dNTPs, and ddNTPs (labeled with fluorescent dyes) to the reaction mixture. DNA synthesis will occur, incorporating both dNTPs and ddNTPs. Occasionally, a ddNTP will be incorporated, causing DNA synthesis to terminate at a specific nucleotide.

4. **Fragment Separation**: Load the reaction mixture onto a denaturing polyacrylamide gel. Apply an electric field using an electrophoresis apparatus. Fragments will migrate based on size, with the terminated fragments stopping at the position of the specific ddNTP.

5. **Detection and Analysis**: Use a DNA sequencing machine with a fluorescence detector to analyze the gel. The detector will read the labeled fragments as they pass through a capillary array, determining the sequence.

Chemical degradation method

Introduced **by Allan Maxam and Walter Gilbert in 1977**, this method relies on treating double-stranded DNA with specific chemicals that cause it to break at precise nucleotide positions. By analyzing the resulting fragments, the sequence can be deduced.

Virtual learning aid

Materials and Equipment

1. Target compound (to be degraded)
2. Chemical reagents (degrading agents)
3. Solvent(s)
4. Reaction vessel (e.g., round-bottom flask)
5. Stirring apparatus (e.g., magnetic stirrer)
6. Heating source (e.g., hot plate)
7. Cooling source (e.g., ice bath)
8. pH meter or indicator paper
9. Glassware (pipettes, beakers, etc.)
10. Safety equipment (lab coat, gloves, safety goggles)
11. Separatory funnel (if required)
12. Analytical instruments (e.g., spectrophotometer, chromatograph)
13. Documentation materials (notebook, pen)

Protocol

1. **Preparation**: Put on appropriate safety gear, including a lab coat, gloves, and safety goggles. Set up the reaction area with proper ventilation. Ensure all equipment is clean and dry.

2. **Weighing and Measuring**: Accurately weigh the target compound to be degraded. Prepare the required amounts of chemical reagents and solvents based on the reaction stoichiometry.

3. **Reaction Setup**: Place the weighed target compound in a clean, dry reaction vessel (round-bottom flask). Add the appropriate solvent to the flask to create a suitable reaction mixture. Attach a condenser (if required) to prevent loss in volatile components.

4. **Stirring and Heating**: Start magnetic stirring to ensure uniform mixing. If heat is required, place the reaction vessel on a hot plate and set the desired temperature. Monitor the reaction progress by observing changes in color, precipitate formation, or other visual cues.

5. **Addition of Degrading Agent**: Gradually add the selected chemical reagent(s) responsible for the degradation reaction. Control the addition rate to maintain a steady reaction rate and prevent over-reaction.

6. **pH Adjustment (if applicable)**: Measure the pH of the reaction mixture using a pH meter or indicator paper. If necessary, adjust the pH using acids or bases to optimize the reaction conditions.

7. **Monitoring and Analysis**: Take periodic samples from the reaction mixture to analyze the progress of degradation. Analyze samples using suitable analytical instruments (e.g., spectrophotometer, chromatography) to determine the extent of degradation.

8. **Cooling and Isolation**: If the reaction generates heat, allow the mixture to cool down to room temperature or below. If required, isolate the degraded product by filtration, extraction, or other appropriate methods.

9. **Safety and Disposal**: Dispose of any chemical waste according to established laboratory protocols. Clean and store equipment properly.

Pyrosequencing method

Developed in **1996 by Mostafa Ronaghi and Pal Nyrén**, this method involves detecting the addition of a deoxynucleotide to a growing DNA strand, which produces a flash of light. This flash indicates the sequence of the DNA.

In the chain termination method, dideoxyribonucleoside triphosphates (ddNTPs) are used, which lack a specific group required for further DNA synthesis. When incorporated into the growing DNA chain, they prevent further elongation, resulting in termination at that point. To sequence a DNA fragment completely, one strand of the double-stranded DNA is separated and used as a template for sequencing. Four types of chain-terminating ddNTPs (ddATP, ddCTP, ddGTP, and ddTTP) are utilized in separate reactions to determine the sequence of the template strand.

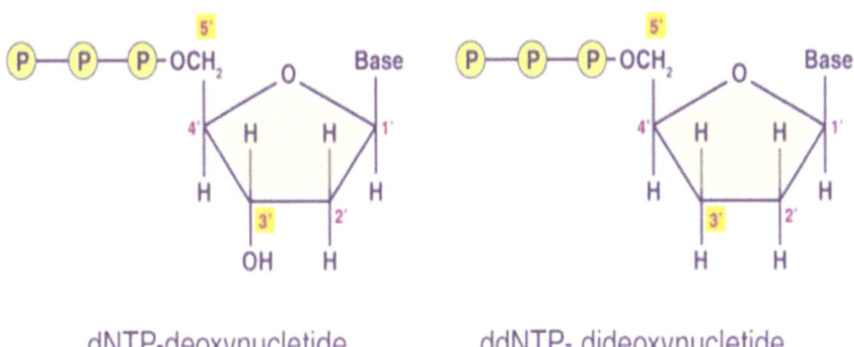

dNTP-deoxynucletide   ddNTP- dideoxynucletide

In the process, each reaction generates a set of DNA copies that end at different positions within the sequence. These resulting products are then separated using electrophoresis in four separate lanes on a polyacrylamide sequencing gel. These gels, which contain denaturing agents like urea and formamide, typically consist of 6-20% polyacrylamide and 7 mol/l urea, facilitating the analysis of DNA sequencing.

Virtual learning aid

Scan It!

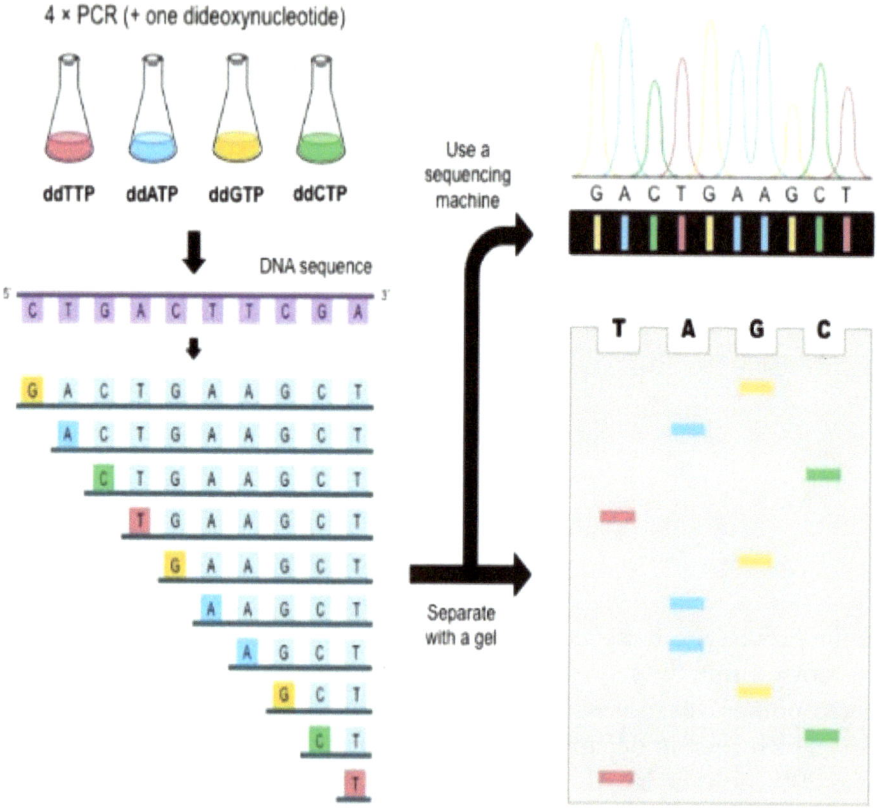

Urea serves a crucial role in DNA analysis by preventing unwanted folding within the DNA strands, which could impact their movement during electrophoresis. This is significant because alterations in shape caused by internal base pairing affect how quickly single-stranded DNA molecules migrate. To enhance separation, DNA sequencing gels are often lengthy (around 100 cm). As DNA strands are extended during synthesis, they are labeled with radioactive or fluorescent markers, and integrated into primers or building blocks. The resulting fragments are separated on the gel, forming distinct bands in lanes. These bands, originating from different positions in the DNA but terminating at specific nucleotides, can be read sequentially from the gel's bottom across lanes to determine the sequence of the newly synthesized DNA strand.

Reagents

1. DNA sample
2. DNA primers (forward and reverse)
3. Enzymes (e.g., DNA polymerase)
4. Nucleotide dispensation reagents (A, T, C, G)
5. Pyrosequencing beads
6. Luciferase enzyme
7. ATP sulfurylase enzyme
8. Adenosine 5'-phosphosulfate (APS)
9. Substrate solution (e.g., luciferin)
10. Enzyme co-factor solution (e.g., magnesium sulfate)

Equipment

1. Thermal cycler
2. Pyrosequencer instrument
3. Computer with Pyrosequencing software
4. Pipettes and pipette tips
5. Microplates and sealing films
6. Reaction tubes or microplates
7. Centrifuge
8. Magnetic stand (for bead separation)
9. Disposable gloves and lab coat

Protocol

1. **PCR Amplification**: Prepare PCR reaction mix with DNA template, primers, DNA polymerase, and nucleotides. Perform thermal cycling to amplify the target DNA region.

2. **PCR Product Cleanup**: Purify the PCR product to remove excess primers and nucleotides using a purification kit.

3. **Annealing of Sequencing Primer**: Anneal the sequencing primer to the purified PCR product.

4. **Pyrosequencing Reaction Setup**: Prepare a mix of Pyrosequencing beads, enzymes (Luciferase and ATP sulfurylase), APS, substrate solution, and enzyme co-factor solution.

5. **Bead Binding**: Bind the DNA template to the Pyrosequencing beads.

6. **Pyrosequencing Instrument Setup**: Load the bead-DNA complex into the pyrosequencer and set up the dispensation reagents (A, T, C, G).

7. **Pyrosequencing Reaction**: Initiate the pyrosequencing run. As nucleotides are added one at a time, the complementary strand is synthesized, and released pyrophosphate generates light.

8. **Data Analysis**: Analyze the pyrosequencing data using specialized software to interpret the sequence and detect variations.

Automated sequencing

The conventional method for DNA sequencing uses radioactive labels and autoradiography to visualize the DNA bands in a gel. However, this approach is not ideal for automation. To achieve automated sequencing, **a real-time detection system is preferred**, where fluorescent dideoxynucleotides are employed. These fluorescent molecules, each emitting a unique color, are used to label DNA fragments. By utilizing different colors for each nucleotide (A, C, G, T), all sequencing reactions can be performed in a single tube and loaded into a single gel lane. The fluorescent detector then distinguishes the colors, allowing direct and automated reading of the sequence as the bands pass by. The output can be presented visually or stored digitally.

Sequencing enzymes

Naturally occurring polymerases have features that are often not optimal for DNA sequencing. The relevant and desirable properties of sequencing polymerases include:

1. **Processivity Level**: Processivity refers to how efficiently an enzyme extends a DNA chain before detaching from a primer-template complex. T7 Sequenase exhibits the highest processivity among current sequencing enzymes, while the Klenow fragment shows the lowest.

2. **Thermostability**: Enzyme stability at high temperatures is crucial for cycle sequencing, a foundation of modern high-throughput sequencing. Taq polymerase, including its variants, remains the primary viable option due to its resistance to heat-induced inactivation.

3. **Nucleotide Analog Incorporation**: Incorporating nucleotide analogs like dye terminators is vital for the dideoxy chain-termination sequencing method.

Similar efficiency in terminating chains using different dye-labeled terminators is essential to produce high-quality sequencing data with consistent peak heights.

4. **Exonuclease Activities**: Enzymes often possess exonuclease activities that proofread DNA by removing RNA primers after replication. However, for sequencing purposes, polymerase variants lacking these activities, such as thermal sequenase, are preferred to avoid unwanted proofreading effects.

Chemical degradation method

The chemical degradation approach involves modifying specific DNA bases and then cleaving the DNA strands to create a set of labeled fragments. Initially, radioactive phosphorus is attached to the 5' end of each strand in double-stranded DNA, which is then separated and analyzed individually. After dividing the single-labeled DNA into four portions, each portion is exposed to chemicals targeting particular bases. This modification occurs at one position along each molecule, ensuring equal modification chances for all bases. Subsequent chemical cleavage breaks phosphodiester bonds at the modified nucleotide positions, yielding labeled fragments. While the cleavage reactions remain largely unchanged from their original design, some new chemical cleavage techniques have been introduced over time. The base modification processes involve specific chemical treatments.

After introducing changes to the DNA bases, the reactive secondary amine known as piperidine is employed to break the sugar-phosphate backbone of the DNA at the locations where the base modifications occurred. This process involves two main steps:
Firstly, chemically altering DNA bases.

Secondly, detaching the modified base from its sugar molecule, and then breaking the phosphodiester bonds that connect the DNA backbone adjacent to the modified base, on both the 5' and 3' sides.

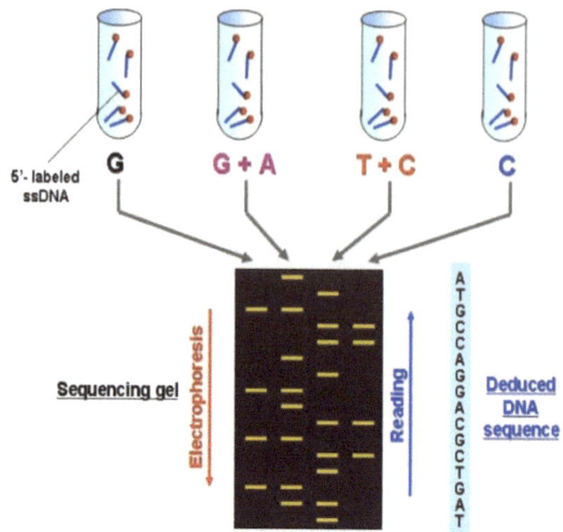

We will use the 'G' cleavage reaction as an example. First, we treat the molecules with dimethyl sulfate, which adds a methyl group to the G nucleotides' purine ring. We only add a small amount to modify about one G per polynucleotide on average. The DNA strands remain intact initially. Cleavage occurs when we introduce piperidine, a second chemical. Piperidine removes the modified purine and cleaves the phosphodiester bonds. This results in a set of cleaved DNA molecules, some labeled and some not. The labeled molecules share one end and have another end determined by the cut sites, indicating G nucleotide positions. These cleaved molecules are then separated using electrophoresis in a polyacrylamide gel, and their sequence is read similarly to chain termination sequencing.

Base(s) Modified	Chemical Treatment	pH	Additional Conditions
G only	Methylation with dimethyl sulfate	8.0	
A+ G	Treatment with piperidine formate	2.0	
C+T	Hydrazine treatment		
C only	Hydrazine treatment		1.5 M NaCl

Chapter 12

Blotting Techniques

Introduction

Blotting is a technique used **to fix sample nucleic acids or proteins onto a solid support**. It involves transferring these molecules from a gel strip onto a specialized membrane, typically made of **nitrocellulose or activated nylon**. This membrane chemically reacts with and immobilizes the nucleic acids/proteins in a pattern similar to the original gel.

Southern blotting

Southern blotting, also known as Southern blot hybridizations, is a method developed **by E. M. Southern in 1975**. It involves transferring and immobilizing DNA fragments that have been separated by gel electrophoresis onto a membrane. After denaturing the DNA fragments on an agarose gel, they are transferred to the membrane for detection and analysis. This technique is commonly used **to study and analyze DNA fragments of varying sizes**.

After performing electrophoresis, DNA fragments are treated with strong alkali and then transferred onto durable nitrocellulose paper or nylon membrane. This transfer ensures stability and prevents diffusion within fragile electrophoretic gels. Fixation of DNA fragments onto the membrane is crucial to prevent detachment. Nitrocellulose paper immobilizes nucleic acids non-covalently after baking, while the nylon membrane achieves covalent binding through UV irradiation. This binding relies on cross-links forming between DNA's T-residues and amino groups on the nylon surface. DNA fragments become immobilized on the membrane, reflecting their size separation during gel electrophoresis. Subsequently, a labeled RNA, single-stranded DNA, or oligodeoxynucleotide solution, known as a probe, is applied to the membrane to detect specific DNA bands. The probe hybridizes with complementary target DNA sequences, and careful conditions are chosen to enhance hybridization while minimizing non-specific binding. After the hybridization process, the membrane is extensively washed to remove non-specifically bound probe.

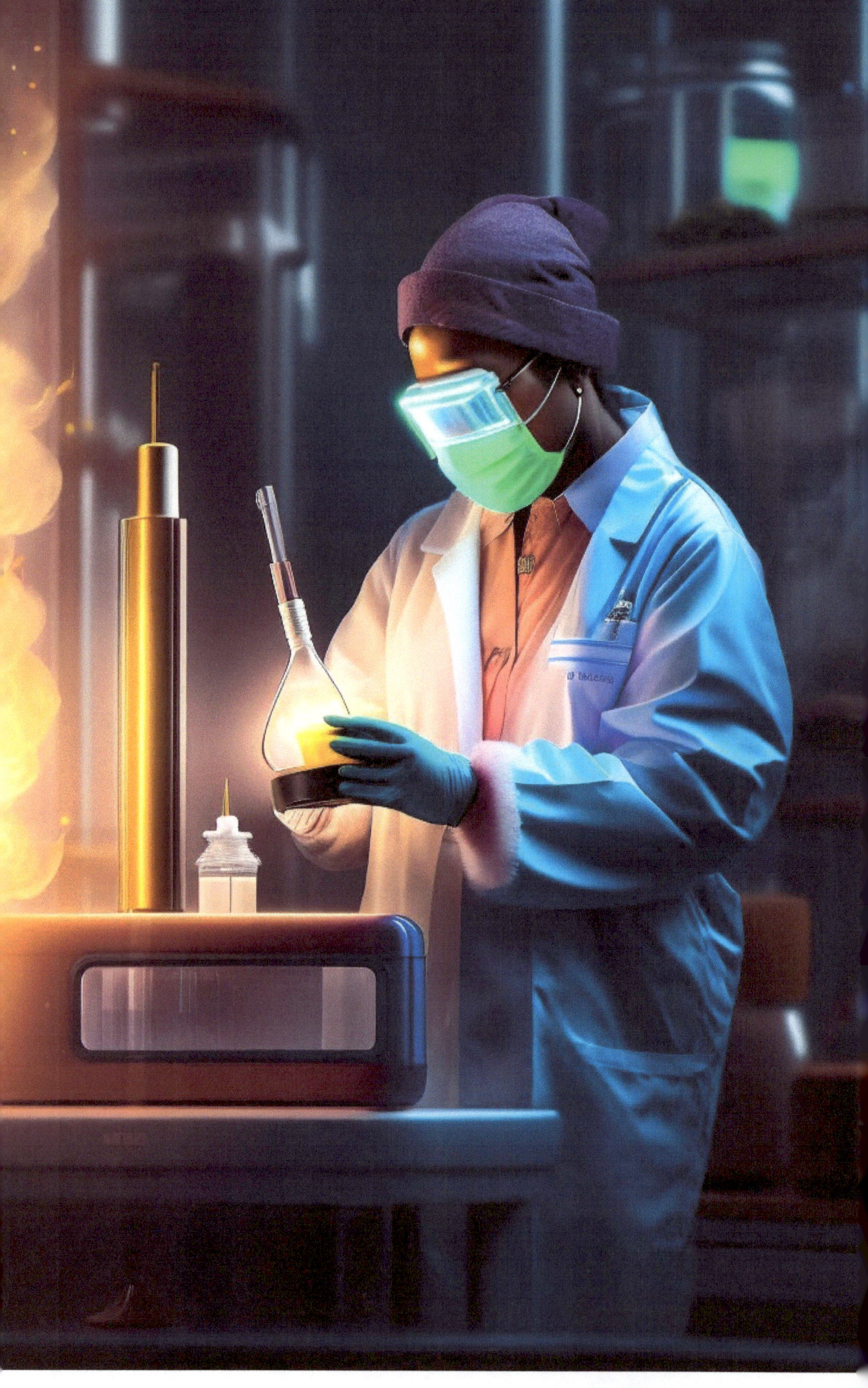

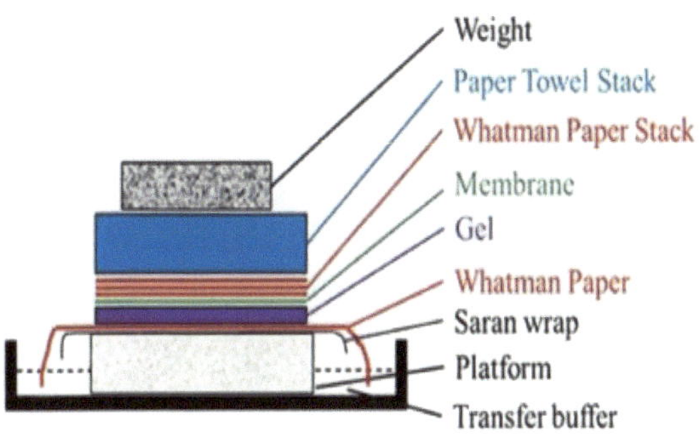

When using a radioisotope-labeled probe, the membrane is exposed to photographic film. Alternatively, for non-isotopically labeled probes like biotin or digoxigenin, the membrane can be treated with a chemiluminescent substrate for probe detection, followed by photographic film exposure. This results in the formation of a distinct band on the film, indicating the probe's position corresponding to the complementary sequence on the membrane.

Virtual learning aid

Materials and Equipment

1. Agarose gel with separated DNA fragments
2. DNA samples for blotting
3. Transfer buffer (20x SSC)
4. Nylon or nitrocellulose membrane
5. Blotting paper or filter paper
6. UV crosslinker (for nylon membrane)
7. Hybridization buffer
8. Radioactively or chemically labeled DNA probe
9. Autoradiography film
10. X-ray developer
11. X-ray film processor

Equipment

1. Gel electrophoresis apparatus
2. Power supply
3. Blotting apparatus (semidry or vacuum)
4. Hybridization oven or incubator
5. Autoradiography equipment

Protocol

1. **Agarose Gel Electrophoresis**: Prepare and run an agarose gel to separate DNA fragments based on size. Load DNA samples onto the gel along with DNA size markers. Run the gel at an appropriate voltage and time to separate DNA fragments.

2. **DNA Transfer**: Cut a piece of nitrocellulose or nylon membrane slightly larger than the gel. Soak the membrane in transfer buffer (20x SSC) for a few minutes. **Assemble the transfer sandwich**: Place the gel, followed by blotting paper, membrane, and more blotting paper. Apply pressure using a semidry or vacuum blotting apparatus for efficient transfer of DNA from the gel to the membrane.

3. **Crosslinking (For Nylon Membrane)**: If using a nylon membrane, UV crosslinks the DNA to the membrane using a UV crosslinker.

4. **Prehybridization and Hybridization**: Prehybridize the membrane in a hybridization buffer to block nonspecific binding. Prepare a labeled DNA probe specific to your target sequence. Hybridize the labeled probe to the membrane by incubating at an appropriate temperature.

5. **Post-Hybridization Washes**: Wash the membrane with stringent washing buffers to remove the unbound probe and reduce the background signal.

6. **Detection**: Place the membrane in contact with an autoradiography film. Allow the film to develop using a suitable developer, or use chemiluminescence if using a chemically labeled probe. Process the developed film in an X-ray film processor.

Northern blotting

Northern blotting, a variation of Southern blotting, focuses on RNA rather than DNA. It gauges RNA abundance and size from specific genes. The process involves **RNA electrophoresis in denaturing conditions to prevent RNA**

base pairing. The separated **RNA is then transferred to a special paper and hybridized with a labeled probe**. Additionally, nylon membranes, categorized as unmodified or positively charged, can be used to capture RNA bands.

In Northern blotting, RNA is analyzed instead of DNA. It helps measure and estimate RNA amounts from genes. The technique involves RNA separation in a gel under denaturing conditions to prevent unwanted base pairing. The isolated RNA is transferred to a paper and paired with a labeled probe. Nylon membranes, available in unmodified or positively charged forms, can also be used to capture RNA bands efficiently.

Virtual learning aid

Materials

1. RNA samples (isolated from cells or tissues)
2. RNA loading buffer
3. Formaldehyde (37% stock solution)
4. RNA size markers (RNA ladder)
5. Nylon membrane (positively charged)
6. Transfer buffer (20x SSC)
7. Denaturing solution (formamide-based)
8. Prehybridization solution
9. Hybridization solution (containing labeled probe)
10. Washing buffers (varying stringencies)
11. Detection system (e.g., chemiluminescence)

Equipment

1. Electrophoresis apparatus and power supply
2. Transfer apparatus (semi-dry or wet)
3. Hybridization oven/incubator
4. Vacuum or hybridization oven for membrane washing
5. UV transilluminator or autoradiography equipment

Protocol

1. **RNA Electrophoresis**: Prepare RNA samples: Mix RNA with loading buffer, denature at **65°C for 10 minutes**, and chill on ice. Load RNA samples and RNA ladder onto denaturing agarose gel. Run the gel at a suitable voltage until the desired separation is achieved. Stain gel with ethidium bromide to visualize RNA bands.

2. **RNA Transfer**: Prepare transfer buffer (20x SSC). Cut nylon membrane to match gel size and wet in transfer buffer. Assemble transfer sandwich: Gel, membrane, filter paper, filter paper, and sponge. Apply pressure evenly. Transfer RNA onto the membrane using a semi-dry or wet transfer apparatus.

3. **Crosslinking RNA to the Membrane**: UV crosslink the RNA on the membrane or bake at **80°C for 2 hours**.

4. **Prehybridization and Hybridization**: Prehybridize the membrane in a prehybridization solution. Prepare labeled probe (radioactive or non-radioactive) and denature. Hybridize the membrane with the labeled probe in a hybridization solution overnight.

5. **Membrane Washing**: Wash the membrane with increasingly stringent washing buffers to remove unbound probes. Allow sufficient time for washing to ensure specific signal detection.

6. **Probe Detection**: For radioactive probes, expose the membrane to X-ray film or autoradiography film. For non-radioactive probes, use a chemiluminescence detection system.

Western Blotting

Western blotting, also known as **immunoblotting**, is a powerful technique used **to detect and quantify specific proteins in a complex mixture**. It involves the **separation of proteins based on their molecular weight using gel electrophoresis**, followed by their transfer onto a membrane and subsequent detection using antibodies.

Fundamental concept

Western blotting exploits the principles of protein separation by size and antigen-antibody recognition. The technique involves three main stages: protein separation by gel electrophoresis, transfer to a membrane (usually nitrocellulose or PVDF), and protein detection using antibodies.

Reagents

1. SDS-PAGE resolving gel solution
2. SDS-PAGE stacking gel solution
3. Protein sample buffer
4. Running buffer
5. Transfer buffer
6. Blocking buffer
7. Primary antibody specific to target protein
8. Secondary antibody conjugated to an enzyme (e.g., HRP)
9. Chemiluminescent substrate

Equipment

1. Electrophoresis apparatus
2. Nitrocellulose/PVDF membrane
3. Power supply
4. Transfer apparatus (semi-dry or wet)
5. Shaker or rocker
6. Blocking tray
7. Western blotting boxes
8. Incubator
9. Chemiluminescence imaging system

Protocol

1. **Protein Electrophoresis**: Prepare resolving gel by mixing appropriate volumes of resolving gel solution, distilled water, and initiator. Likewise, prepare stacking gel. Assemble gel apparatus and degas resolving gel mixture. Load samples along with protein ladder in wells. Run the gel at constant voltage until the dye reaches the bottom.

2. **Protein Transfer**: Cut a piece of nitrocellulose/PVDF membrane to match the size of the gel. Soak the membrane in methanol, followed by transfer buffer. Assemble transfer apparatus and pre-soak blotting pads and membrane in transfer buffer. Carefully transfer proteins from the gel to the membrane using appropriate transfer conditions (voltage, time, and buffer).

3. **Blocking and Antibody Incubation**: Incubate the membrane in a blocking buffer to prevent nonspecific binding. Incubate the membrane with primary antibody diluted in blocking buffer. Wash to remove excess antibody. Incubate with secondary antibody conjugated to an enzyme. Wash again to remove the unbound antibody.

4. **Protein Detection**: Prepare a chemiluminescent substrate and apply it to the membrane. Capture chemiluminescent signal using an imaging system.

Eastern Blotting

Eastern blotting is a powerful molecular biology technique used **to detect and analyze post-translational modifications (PTMs) of proteins**. It is a variation of the well-known Western blotting technique, which is used to detect specific proteins in a sample. In Eastern blotting, the focus shifts to analyzing modifications like glycosylation, phosphorylation, or other chemical changes in proteins.

Fundamental concept

Proteins undergo various modifications that play critical roles in their function and regulation. Eastern blotting allows researchers to visualize these modifications by transferring protein samples from a gel to a solid support (such as a membrane), followed by detection using specific probes or antibodies. This technique provides insights into the diversity of PTMs present in a sample.

Virtual learning aid

Materials and Equipment

1. Protein samples (lysates, cell extracts, etc.)
2. Polyacrylamide gel (SDS-PAGE gel)
3. Nitrocellulose or PVDF membrane
4. Transfer buffer (e.g., Towbin buffer)
5. Power supply for electrophoretic transfer
6. Blocking buffer (e.g., TBS with Tween 20 or BSA)
7. Primary antibodies specific to the PTM of interest
8. Secondary antibodies (conjugated to an enzyme or fluorophore)
9. Detection substrate (chemiluminescent or fluorescent)
10. Imaging system (chemiluminescence or fluorescence imager)
11. Appropriate laboratory safety equipment

Protocol

1. **Prepare the gel**: Perform SDS-PAGE with protein samples to separate proteins based on size. Use appropriate controls and molecular weight markers.

2. **Transfer proteins**: Cut the gel to match the membrane size. Soak the gel and membrane in transfer buffer. Assemble the transfer sandwich: Gel - Nitrocellulose membrane - Filter paper - Sponge. Transfer proteins via electrophoresis (follow manufacturer's instructions) onto the membrane.

3. **Blocking and antibody incubation**: Block the membrane with a blocking buffer to prevent non-specific binding. Incubate the membrane with the primary antibody specific to the PTM of interest. Wash to remove unbound primary antibody.

4. **Secondary antibody and detection**: Incubate the membrane with secondary antibody conjugated to an enzyme or fluorophore. Wash to remove unbound secondary antibodies. Prepare the detection substrate and apply it to the membrane.

5. **Visualization**: Capture chemiluminescent or fluorescent signals using an imaging system. Adjust exposure settings as needed.

6. **Data analysis**: Analyze band intensities using appropriate software. Compare samples and controls for PTM patterns.

Chapter 13

Cell Culture Techniques

Introduction

Since its inception **by Ross Harrison in 1907**, the field of animal cell culture technology has transformed from observation-based science into a versatile research tool. It offers a controlled environment **for studying cell behaviors** like proliferation, differentiation, and product synthesis, contributing to the production of vital medical and research compounds derived from natural biomolecules. Recent advancements have expanded the scope of products, including vaccines, monoclonal antibodies, and therapeutic proteins, achieved through cell culture techniques. Moreover, animal cell cultures serve as valuable models for various research areas, such as toxicity studies, cancer biology, gene therapy, and drug screening. The technique can be implemented using batch or continuous culture methods, including primary explant cultures or suspension cultures. Proficiency in essential techniques is crucial for the successful maintenance of cultured cells, whether primary or immortalized.

Medium preparation for Animal cell culture

Culture media consists of a customized mix of vitamins, salts, glucose, amino acids, and growth factors. These components are tailored to specific cell types, with essential salts like NaCl, KCl, $CaCl_2$, and $MgCl_2$ ensuring osmotic balance and membrane potential. Amino acids, vitamins, and buffering agents like Na_2HPO_4 and NaH_2PO_4 maintain a pH range of 7.2 to 7.4, while glucose and amino acids provide energy and nitrogen. Incorporating antibiotics prevents microbial contamination. Traditional media relied on natural sources like chick embryo extracts and serum, while chemically defined alternatives such as Eagle's basal media, MEM, DMEM, and RPMI 1640 gained popularity due to their precise composition and consistency. Animal cells thrive in liquid mediums with optimized conditions, including pH and osmotic pressure, playing a pivotal role in successful cell growth.

270

Materials

1. Culture medium
2. Adult bovine serum
3. 0.22 μm membrane filter (Millipore)
4. Double distilled water
5. 1-liter measuring cylinder (1000 ml)
6. 100 ml measuring cylinder
7. 1-liter filtration flask
8. Sterile medium storage bottles

Protocol

1. Place 500 ml of sterile double distilled water in a 1000 ml measuring cylinder.

2. Transfer powdered medium into the measuring cylinder and dissolve thoroughly.

3. Add penicillin-G (100 U/ml), streptomycin (100 mg/ml), and gentamycin (50 mg/ml). Adjust volume to 1000 ml with double distilled water.

4. Transfer the medium to a sterile 1-liter flask, observing a pinkish-red color indicating the desired pH.

5. Assemble the filtration unit, and conduct filtration under negative pressure.

6. Prepare 400 ml of medium with 10% adult bovine serum using a 100 ml measuring cylinder. Store in a 500 ml sera lab bottle.

7. Transfer remaining serum-free medium to larger glass bottles.

8. Store medium in a refrigerator and discard the used membrane filter.

Table Representing ingredients of Eagle's minimal essential medium (MEM).

Components	Concentration Range
Amino Acids:	
1. L-Glutamine	292-584 mg/L
Vitamins:	
1. Vitamin B12	Varies
2. Folic Acid	Varies
3. Riboflavin (Vitamin B2)	0.2 mg/L
4. Thiamine (Vitamin B1)	0.5 mg/L
Salts:	
1. Sodium Chloride (NaCl)	6.7 g/L
2. Potassium Chloride (KCl)	0.4 g/L
3. Calcium Chloride (CaCl2)	0.1 g/L
4. Magnesium Sulfate (MgSO4)	0.2 g/L
Glucose	1000-4500 mg/L
Inorganic Salts	Varies
Buffering Agents:	
1. Sodium Phosphate Dibasic (Na2HPO4)	Varies
2. Sodium Phosphate Monobasic (NaH2PO4)	Varies
Phenol Red (pH indicator)	15 mg/L
Fetal Bovine Serum (FBS)	5-10%
Antibiotics:	
1. Penicillin	100 units/mL
2. Streptomycin	100 µg/mL
L-Cystine	24 mg/L
L-Tyrosine	14 mg/L
Choline Chloride	6.2 mg/L
Inositol	18 mg/L

Table representing all ingredients of Dulbecco's modification of Eagle's medium (DMEM).

Component	Concentration
Amino Acids	
1. L-Arginine	0.150 g/L

2. L-Cystine	0.150 g/L
3. L-Histidine	0.150 g/L
4. L-Isoleucine	0.105 g/L
5. L-Leucine	0.105 g/L
6. L-Lysine	0.146 g/L
7. L-Methionine	0.015 g/L
8. L-Phenylalanine	0.105 g/L
9. L-Threonine	0.095 g/L
10. L-Tryptophan	0.015 g/L
11. L-Tyrosine	0.100 g/L
12. L-Valine	0.105 g/L
Vitamins	
1. Choline Chloride	0.0075 g/L
2. Folic Acid	0.001 g/L
3. i-Inositol	0.01 g/L
4. Niacinamide	0.002 g/L
5. Pantothenic Acid	0.01 g/L
6. Pyridoxal Hydrochloride	0.002 g/L
7. Riboflavin	0.001 g/L
8. Thiamine	0.001 g/L
9. Vitamin B12	0.000001 g/L
10. Biotin	0.00003 g/L
11. Ascorbic Acid	0.02 g/L
Salts	
1. Calcium Chloride	0.0011 g/L
2. Magnesium Sulfate	0.098 g/L
3. Potassium Chloride	0.4 g/L
4. Sodium Chloride	6.0 g/L
5. Sodium Phosphate (Dibasic)	1.15 g/L
6. Sodium Phosphate (Monobasic)	0.2 g/L
Other	
1. D-Glucose	1.0 g/L
2. Phenol Red	0.0005 g/L
3. Sodium Pyruvate	0.11 g/L
4. L-Glutamine	0.584 g/L
5. HEPES	0.01 g/L

Primary cell culture

In the realm of scientific inquiry, the development of in vitro techniques for culturing animal cells has emerged as an invaluable tool. By creating controlled environments, researchers gain the ability to delve into the intricate structures and functions of cells. While almost all tissues can be cultured, optimal success often arises from the direct isolation of cells from the tissue. This isolation involves either mechanical dissociation or enzymatic detachment, leading to disaggregated cells that can flourish in a suitable culture medium. The **foundations of primary cell culture** necessitate certain conditions: **seeding a sufficient number of cells per vessel, finely chopping tissues for proper cell dissociation**, and **ensuring the removal of enzymes** used in tissue disaggregation.

Fundamental concept

Primary cultures, often derived from substantial tissue masses, serve as the initial step in cell culture progression. This foundational culture, referred to as the primary cell culture, comprises an array of differentiated cells. Due to lower survival rates, a higher quantity of cells is typically utilized in primary cultures. Embryonic tissue, characterized by easier disaggregation, higher viable cell yields, and rapid proliferation, is commonly favored for primary cell culture. The primary culture can be established by allowing cells to migrate from tissues or through mechanical tissue disaggregation. After isolation, the primary cell culture can be further nurtured in fresh media to establish a secondary culture, extending the investigative journey into the realm of cellular behavior and dynamics.

Virtual learning aid

Requirements

1. **Tissue Source**: Select embryonic tissues or organ samples of interest for cell isolation. The age of the tissue will depend on the experimental requirements.
2. **Sterile Equipment**: Utilize sterile tools, including scissors, forceps, and Petri plates, to maintain aseptic conditions during tissue dissection.
3. **Culture Vessels**: Prepare culture vessels, such as conical flasks or culture bottles, to accommodate the cell suspension.

4. **Media and Supplements**: Prepare a suitable growth medium, which often includes a nutrient-rich base medium like DMEM supplemented with serum (e.g., fetal calf serum) and other essential factors.

5. **Buffer Solution**: Have phosphate-buffered saline (PBS) ready, with appropriate pH and ion concentrations, to wash and suspend cells.

6. **Enzymatic Digestion**: Utilize enzymes like trypsin to dissociate cells from the tissue matrix, enabling their isolation.

Protocol

1. **Tissue Isolation**: Under sterile conditions, dissect the chosen tissue source, removing any unwanted components. Transfer the tissue to a suitable container with PBS.

2. **Tissue Dissociation**: In a controlled environment like a laminar flow hood, mechanically and enzymatically disaggregate the tissue into a cell suspension.

3. **Enzymatic Treatment**: Treat the cell suspension with an appropriate enzyme, such as trypsin, to facilitate cell dissociation. Maintain appropriate temperature and agitation.

4. **Inactivation and Filtration**: Inactivate the enzyme activity by adding a growth medium with serum. Filter the cell suspension to remove debris and obtain a purified cell population.

5. **Centrifugation**: Centrifuge the cell suspension to pellet the cells. Discard the supernatant and resuspend the cell pellet in the growth medium.

6. **Cell Viability and Count**: Assess cell viability using methods like trypan blue exclusion and determine cell concentration using a hemocytometer or automated cell counter.

7. **Seeding and Incubation**: Seed the cells in culture vessels at a suitable density and incubate under optimal conditions (e.g., 37°C and appropriate CO_2 levels) for a defined period.

8. **Media Changes**: Replace the culture medium the next day and monitor cell adherence and viability.

Preparation of established cell lines

Once a primary cell culture is established, certain cells within it undergo survival and proliferation. These proliferated cells are subsequently sub-cultured or passaged. There are **two main types of cell lines**: **finite cell lines** and **continuous cell lines**. The former have a limited lifespan and exhibit consistent behavior, while the latter, following in vitro transformation, give rise to an enduring lineage with a relatively stable phenotype.

Fundamental concept

Following sub-culturing, the primary culture evolves into a cell line, capable to multiple rounds of cultivation. Primary cell lines often undergo prolonged and rapid division, allowing for repeated passaging. Occasionally, specific cells may undergo alterations, leading to distinct morphological characteristics, enhanced growth rates, and increased multiplication. These cells can be cultured extensively and perpetually sub-cultured in a laboratory setting, earning them the designation of established cell lines.

Virtual learning aid

Requirements

1. Monolayer cells from chick embryos
2. Beaker
3. Culture flask
4. Trypsinization flask
5. Growth medium (MEM)
6. Phosphate buffer saline (pH 7.4)
7. Pasteur pipette
8. Trypsin (0.25% in PBSA)
9. Fetal calf serum (FCS)

Protocol

1. **Preparation and Inspection of Culture Flasks**: Examine culture flasks with a fully formed monolayer of cells to ensure they are free from deterioration or contamination.

2. **Removal of Old Medium and Washing**: Discard the old medium and wash the monolayer cells with Phosphate Buffer Saline (PBS) to remove serum traces that may hinder trypsin action.

3. **Trypsinization and Cell Release**: Apply 0.1 ml/cm2 of 0.25% trypsin solution to cover the monolayer, facilitating cell detachment from the substratum within a short period.

4. **Addition of Fresh Medium**: Upon cell detachment, introduce medium (0.2 ml/cm2) containing serum. Utilize a Pasteur pipette to dislodge any cells adhering to the glass surface.

5. **Cell Counting and Dilution**: Perform cell counting using a hemocytometer and dilute the cells to a concentration of approximately 2×104 cells per ml in a 20 ml medium.

6. **Seeding the Flasks**: Seed 5 ml of the cell suspension into two flasks, securely cap them, and place them in an incubator set at 37°C.

7. **Medium Replacement**: Replace the medium with fresh medium every three days to eliminate dead cells along with the old medium, ensuring an optimal growth environment.

8. **Repeat Subculturing**: Repeat the aforementioned steps multiple times, either entirely or partially, to yield established cell lines.

9. **Result**: Throughout the process of repeated subculturing, it's crucial to note that cell lines can undergo significant changes in their characteristics. These changes may include the transition from monolayer growth to forming clumps and irregular cell orientation. Such altered properties are indicating to transformed cell lines, often associated with neoplastic traits.

Maintenance of cell lines

When a cell culture is started, it requires periodic medium changes to maintain its health and viability. This involves passing the cells to ensure a steady supply of nutrients and to prevent the medium from becoming too acidic due to cellular waste accumulation. Typically grown at 37°C, cells form a monolayer in **2 to 5 days**, at which point they need to be transferred to a new flask through feeding, maintaining a consistent regime.

Fundamental concept

Cells adhering to a solid surface need to be detached before subculturing. For finite cell lines, **detachment** is achieved using enzymes like **trypsin**, breaking the protein links between cells and their substrate. **EDTA** is used alongside trypsin to chelate divalent cations such as **calcium** and **magnesium**, crucial for attachment. Care must be taken to neutralize the trypsin/EDTA mixture promptly to prevent cell damage.

Virtual learning aid

Materials

1. Cell line of interest
2. L-glutamine solution (200 mM)
3. Dulbecco's Modified Eagle Medium (DMEM) with 10% Fetal Bovine Serum (FBS)
4. Trypsin-EDTA solution
5. Sterile culture flask
6. Sterile pipettes
7. Isopropanol (70%)
8. Glass bottle
9. Beaker
10. Measuring cylinder

Protocol

1. **Workspace Preparation**: Clean the work surface thoroughly using **70%** isopropanol to ensure a sterile environment for cell culture procedures.

2. **Thawing Trypsin-EDTA**: Prior to usage, thaw the trypsin-EDTA solution according to recommended guidelines.

3. **Disinfection of Equipment**: Disinfect all equipment introduced into the sterile workspace, including media containers and pipettes, using 70% isopropanol.

4. **Medium Removal**: Carefully transfer the culture medium from the cell culture flask into a beaker containing diluted bleach or another suitable disinfectant. Take care to avoid any chlorine splashback or disturbance to the cell layer.

5. **Rinsing the Monolayer**: Gently rinse the cell monolayer with approximately 5 ml of dispersing solution (trypsin-EDTA) per 75 cm^2 flask (or 2 ml per 25 cm^2 flask). Ensure thorough coverage of internal culture vessel surfaces. Transfer the used dispersing solution to the beaker with bleach.

6. **Introduction of Fresh Dispersing Solution**: Add approximately 2 to 3 ml of fresh dispersing solution per 75 cm^2 flask (or 1 to 2 ml per 25 cm^2 flask) using a sterile pipette. Ensure even distribution of the solution across the entire cell sheet. Gently rock the culture vessel for 3 to 5 minutes to facilitate contact between the dispersing solution and the cell sheet, promoting cell detachment.

7. **Gentle Dislodging of Cells**: When the cell sheet is adequately loosened by the dispersing solution, gently strike the flask on your palm to dislodge the cells from the surface.

8. **Breaking Up Cell Aggregates**: Add 10 ml of freshly prepared DMEM medium to the flask. Slowly pipette the medium multiple times to break up any cell aggregates and ensure a homogenous cell suspension.

9. **Flask Labeling and Incubation**: Label the flask(s) with the initials of the cell line, passage number, and date. Place the flask(s) in the incubator set to the optimal growth temperature (37°C) to allow cells to form a monolayer without changing the medium.

10. **Results**: Incubate the cells under appropriate conditions to promote their growth and formation of a confluent cell monolayer.

Preservation of cell lines

Cell lines, while invaluable in research, are susceptible to variations and challenges like genetic instability, senescence, phenotypic changes, microbial contamination, and limited lifespan in finite cell lines. Replacing these lines is expensive and time-consuming. Preserving cell lines is essential to maintain their genetic and phenotypic stability and safeguard against microbial contamination. Storing cells in **liquid nitrogen** at **-196°C** allows them to remain viable indefinitely. The success of this preservation method depends on the conditions during freezing. The freezing process involves gradual cooling from room temperature to **-80°C** at a controlled rate of around **1°C to 3°C per minute**. This controlled cooling rate can be achieved using specialized equipment like a **freezing chamber** or a **cryofreezing container** filled **with isopropanol**, which provides the precise cooling rate needed for successful cell **cryopreservation**. After this initial stage, the **cryogenic vials** are transferred to a liquid nitrogen storage tank, where they can be preserved until needed. Certain factors, such as **ice crystal formation**, **accumulation of toxic metabolites**, **pH levels**, and **cellular precipitates**, must be carefully managed during preservation. To mitigate potential damage during preservation, cryopreservatives like **dimethylsulphoxide (DMSO)**, **glycerol**, and **polyvinylpyrrolidone (PVP)** are employed.

Fundamental concept

Cryopreservation, the technique of preserving cells by introducing cryoprotective agents like DMSO and glycerol and storing them in liquid nitrogen, is a critical method. During cryopreservation, cellular metabolic processes slow down, ensuring cell viability during subsequent divisions. Freezing essentially removes water from cells. Achieving optimal freezing conditions for maximum viability upon thawing involves minimizing intracellular ice crystal formation and reducing damage caused by high-concentration solutes formed when water inside cells freezes. This is achieved by:

1. **Gradual Freezing**: Cells are frozen slowly to allow water to leave the cells, discouraging ice crystal growth.

2. **Hydrophilic Cryoprotectants**: Cryoprotective agents with water-loving properties are used to shield cells from damage during freezing.

3. **Low-Temperature Storage**: Cells are stored at extremely low temperatures to minimize the adverse effects of high salt concentrations on protein structure.

4. **Rapid Thawing**: Thawing is performed quickly to minimize ice crystal growth and its detrimental effects on cell integrity.

Virtual learning aid

Scan It!

Materials

1. Healthy finite cell line suspended in Phosphate-Buffered Saline (PBS) at pH 7.4
2. Trypsin solution (0.25%)
3. Growth medium
4. Cryoprotectant - Dimethyl Sulfoxide (DMSO)
5. Cryovials
6. Hemocytometer
7. Insulated container for freezing: Polystyrene box

Protocol

1. **Cell Selection**: Begin by choosing a healthy cell line that exhibits robust growth and is free from any signs of contamination.

2. **Cell Preparation**: Treat the cells with **5 ml** of growth medium containing **20%** Fetal Calf Serum (FCS) to initiate trypsinization.

3. **Cell Counting**: Utilize a hemocytometer to accurately count the cells in the suspension. Adjust the cell density to 2×10^6 **cells/ml**.

4. **Cryoprotectant Addition**: Gradually introduce Dimethyl Sulfoxide (DMSO) to the cell suspension, reaching a final concentration of **10-20%**. This DMSO-treated solution will serve as the freezing medium.

5. **Dilution and Mixing**: Combine the cell suspension with the freezing medium at a **1:1** ratio, yielding a final cell concentration of approximately 1×10^6 **cells/ml** with a DMSO concentration of **10%**. Thoroughly mix the components to ensure uniform distribution.

6. **Cryovial Filling**: Transfer the well-mixed cell suspension into appropriately labeled cryovials (CRYO tubes ampoules), ensuring each cryovial is securely sealed.

7. **Cooling at Controlled Rate**: Place the cryovials in an insulated container, such as a polystyrene box, and initiate controlled cooling at a rate of **1°C per minute**. This gradual cooling process is essential to prevent cell damage.

8. **Overnight Storage at -70°C**: Once the cooling rate reaches **-70°C**, transfer the cryovials to a **-70°C** freezer and allow them to remain overnight. This intermediate step prepares the cells for the subsequent ultra-low temperature storage.

9. **Transfer to Liquid Nitrogen Freezer**: Promptly transfer the cryovials from the **-70°C** freezer to a liquid nitrogen (**-196°C**) freezer. This transfer must occur swiftly (within **< 2 minutes**) to prevent cell deterioration due to rapid warming. The temperature of **-50°C** should not be exceeded during this process, as it could compromise cell viability.

10. **Freezer Index Entry**: Upon safely placing the cryovials in the liquid nitrogen freezer, document pertinent information in the freezer index to ensure accurate tracking and future retrieval.

Table representing types of selected media for bacterial growth

Media Name	Composition	Application
Nutrient Agar	Peptone, beef extract, agar	General-purpose medium for the cultivation of a wide range of bacteria.
LB (Luria-Bertani)	Tryptone, yeast extract, sodium chloride, agar (optional)	Commonly used medium for routine cultivation, propagation, and maintenance of *Escherichia coli*.
MacConkey Agar	Peptone, lactose, bile salts, neutral red, crystal violet, agar	Selects for Gram-negative bacteria and differentiates lactose fermenters from non-fermenters.
Blood Agar	Tryptic digest of casein, yeast extract, sheep blood, agar	Used to cultivate fastidious bacteria and to distinguish between different types of hemolysis.

Mannitol Salt Agar	Peptone, beef extract, mannitol, salt (7.5% NaCl), phenol red, agar	Selective and differential medium for the isolation and differentiation of Staphylococcus species.
EMB Agar	Peptone, lactose, eosin Y, methylene blue, agar	Selective for Gram-negative bacteria and differentiates between lactose fermenters and non-fermenters.
Cetrimide Agar	Peptone, cetrimide, agar	Selective for *Pseudomonas aeruginosa* and used in clinical and environmental samples.
MRS Agar	Peptone, meat extract, yeast extract, dextrose, dipotassium hydrogen phosphate, magnesium sulfate, agar	Specifically used for the isolation and cultivation of lactic acid bacteria.
TCBS Agar	Peptone, yeast extract, sodium thiosulfate, sodium citrate, sucrose, ferric citrate, bromothymol blue, agar	Selective for Vibrio species, particularly *Vibrio cholerae*.

Table representing types of selected growth media for yeast.

Media Name	Composition	Application
YPD (Yeast Peptone Dextrose)	Peptone, yeast extract, glucose	General-purpose medium for yeast cultivation
SD (Synthetic Dextrose)	Ammonium sulfate, glucose, synthetic amino acids, vitamins	Defined medium for specific growth conditions
YNB (Yeast Nitrogen Base)	Ammonium sulfate, glucose, salts, vitamins, trace elements	Used for selection and maintenance of auxotrophic mutants

SC (Synthetic Complete)	Synthetic Dextrose medium supplemented with amino acids, vitamins, and nucleotide bases	Selective medium for yeast genetic experiments
Minimal medium	Inorganic salts, glucose, ammonium salts, vitamins, trace metals	Studying minimal nutritional requirements
YPGal (Yeast Peptone Galactose)	Peptone, yeast extract, galactose	Inducing specific metabolic pathways
YPRaf (Yeast Peptone Raffinose)	Peptone, yeast extract, raffinose	Inducing specific metabolic pathways
YPLac (Yeast Peptone Lactate)	Peptone, yeast extract, lactate	Inducing specific metabolic pathways
SC-Leu	Synthetic Complete medium lacking leucine	Selective medium for leucine auxotrophic strains
SC-His	Synthetic Complete medium lacking histidine	Selective medium for histidine auxotrophic strains
SC-Ura	Synthetic Complete medium lacking uracil	Selective medium for uracil auxotrophic strains
SC-Trp	Synthetic Complete medium lacking tryptophan	Selective medium for tryptophan auxotrophic strains
SC-Ade	Synthetic Complete medium lacking adenine	Selective medium for adenine auxotrophic strains
SC-His-Leu	Synthetic Complete medium lacking histidine and leucine	Selective medium for double auxotrophic strains
SC-Ura-His	Synthetic Complete medium lacking uracil and histidine	Selective medium for double auxotrophic strains
SC-Trp-Leu	Synthetic Complete medium lacking tryptophan and leucine	Selective medium for double auxotrophic strains
SC-Ade-His	Synthetic Complete medium lacking adenine and histidine	Selective medium for double auxotrophic strains

Table Representing the recipe of mammalian cell culture complete media.

Components	Composition
Dulbecco's Modified Eagle Medium (DMEM)	500 ml
Fetal Bovine Serum (FBS)	10-15% (50-75 ml)
L-Glutamine	2 mM (2 ml of 200 mM stock)
Penicillin-Streptomycin (Pen-Strep)	1% (5 ml of 100x stock)
Sodium Bicarbonate (NaHCO3)	3.7 g (for 500 ml DMEM)
Optional Growth Factors	As needed
Other Supplements	As needed
Phenol Red (pH Indicator)	As needed (usually yellow or red

Chapter 14

Designing Experiments

Introduction

This chapter is meticulously crafted to ignite passion and impart practical knowledge to students and young researchers. Here, I dive into the fictional scientific challenges, designed to stretch the boundaries of biotechnology and the scope of this book too. These hypothetical scenarios offer a dynamic platform for understanding and applying concepts in real-world contexts. This chapter is a catalyst for refined scientific thinking, serving as an inspiration to secure good research funding and make substantial contributions to the realms of biotechnology.

Scenario-1

Emily, a dedicated Ph.D. candidate at a leading research institution, has been assigned a challenging project **to identify and purify specific proteins from a complex mixture extracted from a rare plant species**. These proteins have potential therapeutic applications. Emily's task is to devise a strategy that combines both electrophoresis and chromatography techniques to efficiently separate and analyze the mixture.

Scientific approach

Emily knows that a combination of electrophoresis and chromatography techniques will provide her with a powerful toolset for tackling this complex challenge.

1. **Electrophoresis (SDS-PAGE)**: Emily starts by using Sodium Dodecyl Sulfate Polyacrylamide Gel Electrophoresis (SDS-PAGE) to separate the protein mixture based on molecular weight. SDS denatures the proteins and imparts a negative charge relative to their mass, allowing them to migrate

through the gel. This technique gives her a clear view of the protein bands in the mixture.

2. **Chromatography (Size Exclusion)**: Emily proceeds to use Size Exclusion Chromatography (SEC), also known as gel filtration chromatography. She loads the protein mixture onto a column packed with porous beads. Larger proteins elute first because they are not able to enter the small pores and thus have a shorter path through the column. This separation process helps her isolate the target proteins from the rest of the mixture.

3. **Chromatography (Ion Exchange)**: Next, Emily employs Ion Exchange Chromatography. This technique exploits the charge differences among proteins. She loads the separated protein mixture onto a column containing a resin with charged groups. Positively charged proteins will bind to negatively charged groups on the resin, allowing Emily to elute them separately. This step further purifies the target proteins.

Conclusion

By combining electrophoresis and chromatography techniques, Emily successfully identifies and purifies the target proteins from the complex mixture. Her well-designed scientific approach enabled her to isolate the proteins for further analysis and potential applications in the field of medicine.

Scenario-2

Meet Sarah, a dedicated Ph.D. candidate at a leading research institution. Her project revolves around **developing innovative drug delivery systems using nanoparticles**. She's particularly focused on employing Spectroscopy and Microscopy techniques to characterize the nanoparticles she's been developing. Her goal is to ensure these nanoparticles can effectively transport therapeutic compounds to targeted cells in the body.

Scientific approach

1. **Spectroscopy Analysis**: Sarah begins by using various spectroscopy techniques to analyze the nanoparticles' properties. UV-Vis Spectroscopy provides information about the absorbance and extinction characteristics of the nanoparticles. This helps her determine their size, shape, and surface characteristics.

2. **Fluorescence Spectroscopy**: To study the interaction between nanoparticles and drug molecules, Sarah utilizes fluorescence spectroscopy. She attaches fluorescent tags to the drug molecules and observes how their fluorescence

changes upon binding to the nanoparticles. This provides insights into the binding affinity and stability of the drug-nanoparticle complex.

3. **Microscopy Visualization**: Next, Sarah employs microscopy techniques for visualizing the nanoparticles at a micro and nanoscale. Using Transmission Electron Microscopy (TEM), she obtains high-resolution images of individual nanoparticles. This allows her to confirm the nanoparticles' size and morphology.

4. **Scanning Electron Microscopy (SEM)**: By utilizing SEM, Sarah gains insights into the surface topography of the nanoparticles. This information is crucial as it affects their interaction with biological entities.

5. **Atomic Force Microscopy (AFM)**: Sarah also employs AFM to generate three-dimensional images of the nanoparticles. AFM helps her understand the mechanical properties of nanoparticles, which can impact their behavior within the body.

6. **Data Integration**: Sarah integrates the data obtained from spectroscopy and microscopy techniques to create a comprehensive profile of the nanoparticles. This profile includes size distribution, surface characteristics, binding affinity with drugs, and mechanical properties.

Conclusion

Based on the combined analysis, Sarah concludes that the nanoparticles exhibit ideal characteristics for efficient drug delivery. The UV-Vis and fluorescence spectroscopy provide crucial information about size and drug binding. Microscopy techniques offer visual confirmation of nanoparticle morphology and surface features. By blending Spectroscopy and Microscopy, Sarah successfully characterizes her nanoparticles for drug delivery. Her findings contribute to the development of advanced drug delivery systems, potentially revolutionizing medical treatments.

Scenario-3

Sarah, a dedicated PhD student at a cutting-edge research institute, is tasked with **isolating a rare population of stem cells from a mixed cell culture**. The challenge lies in the fact that these stem cells have similar properties to other cell types present in the culture. The goal is to develop a method to selectively isolate and enrich the stem cell population for further study.

Scientific approach

To address this challenge, Sarah designs a multi-step scientific approach that combines centrifugation and cell culture techniques.

1. **Preparation of Cell Culture**: Sarah starts by maintaining the mixed cell culture and optimizing its conditions to ensure the cells are in their best state for experimentation.

2. **Density Gradient Centrifugation**: Sarah uses density gradient centrifugation, a technique where cells are layered on top of a density gradient medium in a centrifuge tube. As the tube spins, cells move through the gradient and separate based on their density. This allows Sarah to isolate cells of specific densities, which might correspond to the stem cell population.

3. **Cell Viability and Identification**: After centrifugation, Sarah carefully collects the different cell fractions and assesses their viability. She then performs flow cytometry, a powerful technique that allows her to identify and sort cells based on their surface markers. By comparing the marker profile of the different fractions to known stem cell markers, Sarah can determine which fraction contains the enriched stem cell population.

4. **Secondary Culture Enrichment**: Sarah takes the fraction containing the potential stem cells and establishes a secondary cell culture. This culture is designed to further enrich the stem cell population. By providing the right growth factors and conditions, she encourages the stem cells to thrive while other cell types might not adapt as effectively.

5. **Validation and Characterization**: Once the enriched stem cell population is obtained, Sarah conducts rigorous validation tests to confirm their identity and properties. This involves assessing their differentiation potential, gene expression profile, and functionality.

Conclusion

By combining centrifugation to separate cells based on density and cell culture techniques to further enrich the desired population, Sarah successfully overcomes the challenge of isolating and enriching specific stem cells from a mixed culture. Her innovative approach opens new avenues for studying these valuable stem cells in-depth.

Scenario-4

Sophia is a PhD student at a prestigious research institute, specializing in molecular diagnostics. She is working on developing a diagnostic assay **to detect**

a specific protein marker associated with an early-stage autoimmune disease. This disease is difficult to diagnose using conventional methods, and her research could have a significant impact on improving patient outcomes.

Scientific challenge

Sophia faces the challenge of accurately detecting a low concentration of the disease-specific protein marker in patient serum samples. The protein is present in trace amounts and is easily masked by other proteins and compounds present in the samples.

Scientific approach

Sophia devises a multi-step approach utilizing Immunotechniques and Blotting Techniques to overcome this challenge:

1. **Sample Preparation**: Sophia starts by collecting serum samples from both healthy individuals and those suspected of having the autoimmune disease. She carefully processes the samples to isolate the proteins, ensuring minimal degradation or denaturation.

2. **SDS-PAGE Electrophoresis**: Sophia employs SDS-PAGE (Sodium Dodecyl Sulfate Polyacrylamide Gel Electrophoresis) to separate the proteins in the serum samples based on their molecular weight. This technique helps in concentrating the target protein and separating it from other proteins that could interfere with its detection.

3. **Western Blotting**: After electrophoresis, Sophia transfers the separated proteins from the gel to a membrane in a process known as Western Blotting or Immunoblotting. She selects a membrane with high binding capacity to ensure efficient protein immobilization.

4. **Blocking and Antibody Incubation**: Sophia blocks the membrane to prevent the non-specific binding of antibodies. She then incubates the membrane with a specific primary antibody that recognizes the disease-specific protein marker.

5. **Secondary Antibody and Detection**: Sophia uses a secondary antibody labeled with an enzyme or fluorescent marker that binds to the primary antibody. This secondary antibody provides a signal that can be visualized and quantified.

6. **Signal Visualization and Analysis**: By using a substrate that reacts with the enzyme on the secondary antibody, Sophia generates a detectable signal. She captures the signal using specialized imaging equipment or other detection methods, depending on the type of secondary antibody used. Quantitative

analysis of the signal intensity allows her to determine the concentration of the disease-specific protein marker in the patient samples.

Conclusion

Sophia's innovative approach combining Immunotechniques and Blotting Techniques enables her to detect the disease-specific protein marker even in low concentrations within patient serum samples. Her research contributes to the development of a reliable diagnostic assay that can potentially revolutionize the early detection of autoimmune disease, leading to timely intervention and improved patient care.

Scenario-5

Meet Sarah, a passionate graduate student at a renowned research institute. She's been assigned the challenging task of **purifying a recombinant protein for functional analysis**. The protein, a potential candidate for a groundbreaking medical treatment, is a complex one, and traditional purification methods haven't yielded the desired results.

Scientific challenge

Sarah's challenge revolves around achieving a high-purity sample of the recombinant protein. The protein is required in sufficient quantity and quality for functional assays, but it is proving difficult due to its interactions with impurities and other cellular components. The protein is delicate and sensitive, making its purification even trickier.

Scientific Approach

To tackle this challenge, Sarah first considers a combination of chromatography techniques. She decides to use a multi-step purification process to increase her chances of success.

1. **Ion Exchange Chromatography**: Sarah's first step involves using ion exchange chromatography. She chooses an anion exchange column that can capture the protein based on its charge. By carefully controlling the pH and ionic strength of the mobile phase, she can separate the protein from the other impurities that have different charge characteristics.

2. **Affinity Chromatography**: For the second step, Sarah employs affinity chromatography. She utilizes a column with a matrix that has a ligand with a high affinity for her protein of interest. This allows her to selectively bind the protein while washing away undesired components. Eluting the protein requires altering conditions to weaken the protein-ligand interaction.

3. **Size Exclusion Chromatography**: As a final step, Sarah opts for size exclusion chromatography. This technique helps her separate molecules based on their size. She chooses a gel filtration column that allows smaller molecules to enter the pores of the matrix, while the larger protein is excluded. This step provides the finishing touch to the purification process, ensuring that only the pure protein remains.

Conclusion

Through her meticulous scientific approach, Sarah successfully purifies the recombinant protein for functional analysis. The combination of ion exchange, affinity, and size exclusion chromatography helps her overcome the challenges of purification and isolation. With her high-purity protein sample in hand, Sarah can now move forward with the crucial functional assays that will determine its suitability for medical applications.

Scenario-6

Meet Emily, a dedicated PhD candidate at a prestigious research institution. She has been investigating a rare genetic disorder that seems to have a complex genetic basis. Her goal is **to identify the specific DNA sequence variations responsible for the genetic disorder**. Emily has isolated DNA samples from both affected and unaffected individuals, but there's a challenge: the disorder seems to involve a relatively long stretch of DNA, and traditional PCR techniques have yielded ambiguous results.

Scientific Approach

Emily decided to utilize a combination of electrophoresis and sequencing techniques to address this challenge.

1. **Electrophoresis Step**: Emily first employs gel electrophoresis to separate the DNA fragments based on size. She uses agarose gel electrophoresis to visualize the general distribution of DNA fragments in the samples. This gives her an initial understanding of the complexity of the genetic variations involved.

2. **Sequencing Step**: However, Emily realizes that she needs more precise information about the specific DNA sequence variations. She decides to move on to sequencing techniques. Specifically, she chooses Sanger sequencing due to its accuracy in determining the DNA sequence of a relatively short stretch.

3. **Data Analysis**: Once the sequences are obtained, Emily faces the challenge of aligning and comparing the sequences of affected individuals with those unaffected individuals. She uses specialized software to identify any variations or mutations present in the affected group but absent in the unaffected group.

4. **Verification**: Emily identifies a potential candidate sequence variation that is significantly more common in the affected individuals. To validate this finding, she performs additional experiments using other sequencing techniques like Next-Generation Sequencing (NGS), which can provide more comprehensive data across a wider range of DNA fragments.

5. **Functional Analysis**: With a potentially disease-associated DNA sequence variation in hand, Emily collaborates with her colleagues in the lab to conduct functional analyses to understand the impact of this variation on the protein it encodes. This involves molecular biology techniques such as cloning the variant DNA sequence and expressing the mutated protein to observe any changes in function.

Conclusion

Through the combined approach of gel electrophoresis, Sanger sequencing, NGS, and functional analysis, Emily successfully identified a novel DNA sequence variation associated with the rare genetic disorder. Her findings contribute to a deeper understanding of the disorder's genetic basis, paving the way for potential therapeutic interventions in the future.

Scenario-7

Sarah, a dedicated Ph.D. candidate at a prestigious research institution, is engrossed in her studies of cellular dynamics. Her research focuses on **understanding how different culture conditions affect behavior of cancer cells**. She's particularly interested in observing the **interactions between cancer cells and their surrounding microenvironment**.

Scientific challenge

Sarah's challenge lies in capturing these dynamic cellular processes in real time while maintaining a controlled environment. She needs to design experiments that enable her to track changes in cell behavior under varying conditions, such as changes in nutrient availability, growth factors, and oxygen levels.

Scientific approach

1. **Cell Line Selection**: Sarah chooses a relevant cancer cell line and optimizes its growth conditions to ensure consistent behavior.

2. **Microscopy Setup**: She sets up a specialized live-cell imaging system equipped with advanced fluorescence microscopy. The system maintains the cells at a stable temperature, humidity, and CO_2 concentration.

3. **Fluorescent Labels**: Sarah uses fluorescent probes that specifically label cellular structures and molecular processes she wants to observe. These labels allow her to visualize changes in real time.

4. **Culture Conditions**: Sarah prepares multiple culture conditions, each representing a distinct microenvironment factor (e.g., nutrient-rich, nutrient-deprived, hypoxic). She introduces the labeled cancer cells into each condition.

5. **Time-Lapse Imaging**: Sarah captures time-lapse images of the cells over an extended period. She takes images at regular intervals to track changes in cell morphology, migration, and signaling.

6. **Image Processing**: Sarah utilizes advanced image processing software to analyze the acquired images. She uses algorithms to quantify cell movement, proliferation rates, and changes in fluorescent signal intensity.

7. **Comparative Analysis**: Sarah compares the behavior of cells under different culture conditions. She identifies trends and patterns in cellular dynamics and establishes correlations between environmental factors and cell responses.

8. **Statistical Analysis**: She employs statistical tools to validate the significance of her findings. This helps her draw meaningful conclusions from the data.

Conclusion

Through her meticulous experimental design and comprehensive data analysis, Sarah successfully uncovers how various culture conditions impact cancer cell behavior. Her research provides valuable insights into potential therapeutic targets and strategies for manipulating the microenvironment to influence cancer cell dynamics.

Scenario-8

Emma is a PhD student at a prestigious research institute, dedicated to advancing vaccine development. Her current project involves **quantifying antigen-antibody interactions** to optimize a crucial component of a potential vaccine. The research team has identified a promising antigen that could serve as a target for immunity development. However, **determining the precise binding affinity between the antigen and the antibodies** is essential to ensure the vaccine's efficacy. Emma's challenge is to design an experiment that accurately quantifies these interactions.

Scientific Approach

Emma begins by utilizing a combination of Spectroscopy and Immunotechniques to tackle this complex task. She decides to employ Surface Plasmon Resonance (SPR) spectroscopy, a powerful technique for studying biomolecular interactions in real time. Emma immobilizes the antigen onto the SPR sensor surface and functionalizes another surface with the corresponding antibodies.

1. **Preparation of Antigen and Antibodies**: Emma purifies the antigens and antibodies separately to ensure their integrity and purity. This step is crucial to avoid any potential interference that might affect the binding kinetics.

2. **SPR Experiment Setup**: Emma sets up the SPR instrument and establishes a flow system. She introduces a solution of antibodies over the antigen-coated surface and monitors changes in the refractive index caused by binding events.

Virtual learning aid

3. **Data Collection**: As the antibodies bind to the antigen, Emma collects real-time data on binding kinetics, including association and dissociation rates. The SPR instrument provides a sensorgram, allowing her to analyze the interactions quantitatively.

4. **Data Analysis**: Emma employs specialized software to analyze the sensorgram data. By fitting the data to appropriate kinetic models, she calculates the binding affinity (Kd) between the antigen and antibodies. This information is crucial for understanding the strength of the interactions.

5. **Validation and Optimization**: To ensure the accuracy of her results, Emma performs multiple replicates and controls. She also conducts experiments using different concentrations of antibodies to observe the impact on binding kinetics. This step helps her validate her findings and optimize experimental conditions.

6. **Interpretation and Application**: Emma's results indicate the precise binding affinity between the antigen and antibodies. This information is crucial for vaccine development, as it allows the research team to optimize antigen presentation and antibody response in the final vaccine formulation.

Conclusion

In this scenario, Emma's skillful integration of Spectroscopy (SPR) and Immunotechniques enables her to quantitatively analyze the antigen-antibody interactions. Her scientific approach provides valuable insights for the development of an effective vaccine against the targeted pathogen.

Scenario-9

Emily, a dedicated PhD student at a prestigious research institution, is working on a project involving the **identification of a specific protein in a complex biological sample**. Her goal is to detect this protein using Western blotting, a technique that allows the detection of specific proteins within a sample based on their molecular weight. However, Emily is facing a challenge the protein interest is present in very low abundance, making it difficult to detect using standard Western blotting procedures.

Scientific approach

Emily devises a scientific approach to overcome the challenge of detecting the low-abundance protein in her sample:

1. **Centrifugation for Protein Concentration**: Emily decides to use centrifugation to concentrate the protein of interest from her sample. She carefully selects a centrifugation protocol that will separate the protein from the rest of the sample components based on differences in density. By spinning the sample at a specific speed and duration, she manages to pellet the protein and reduce the volume of the sample while retaining the protein of interest.

2. **Blotting Techniques - Western Blot**: With the concentrated protein sample in hand, Emily proceeds to perform a Western blot. She separates the proteins based on their molecular weight using SDS-PAGE (Sodium Dodecyl Sulfate-Polyacrylamide Gel Electrophoresis) and transfers them onto a membrane through electrophoretic blotting. However, instead of using the conventional blotting buffer, she opts for a buffer that enhances the transfer efficiency of low-abundance proteins.

3. **Blocking and Antibody Incubation**: After transferring the proteins to the membrane, Emily blocks non-specific binding sites on the membrane using a blocking solution. She then incubates the membrane with a primary antibody

that specifically targets the protein of interest. Since the protein is now concentrated, the chances of antibody binding and detection are significantly increased.

4. **Enhanced Detection**: Emily uses a sensitive detection method, such as chemiluminescence, to visualize the bound antibody. The concentrated protein yields a stronger signal on the membrane, making it easier to detect even in low-abundance conditions.

Conclusion

Through her strategic combination of centrifugation and Western blotting techniques, Emily successfully concentrates the low-abundance protein from her complex sample. This innovative approach allows her to achieve the sensitivity needed for accurate detection and analysis. Emily's scientific approach showcases the importance of combining different techniques to overcome experimental challenges and push the boundaries of protein analysis in research.

Scenario-10

Emma, a PhD student at a renowned research institute, is working on unraveling the genetic basis of **a rare hereditary disorder** that has eluded scientists for years. She believes that **a specific gene is responsible for the disorder**, but she needs to identify and sequence it to confirm her hypothesis. The **gene is suspected to be present in extremely low amounts in the patients' DNA samples**. Emma faces the challenge of **isolating and purifying the gene from a complex mixture of genetic material**.

Scientific Approach

Emma employs a multi-step scientific approach to overcome the challenge to isolating and sequencing the gene of interest:

1. **DNA Extraction and Purification**: Emma starts by extracting DNA from patient samples using well-established techniques. To increase the purity of the DNA, she employs column chromatography. This technique utilizes a solid matrix with specific binding properties that allow the DNA to selectively bind while impurities are washed away. This results in highly purified DNA suitable for sequencing.

2. **Polymerase Chain Reaction (PCR)**: To amplify the specific gene region, Emma employs PCR. This technique creates millions of copies of the target gene segment, making it easier to work with and increasing the chances of successful sequencing.

3. **Gel Electrophoresis**: Emma verifies the success of her PCR amplification by running a gel electrophoresis experiment. This technique separates DNA fragments based on size, confirming the presence of the amplified gene segment.

4. **DNA Sequencing**: Emma chooses Sanger sequencing for its accuracy in reading DNA sequences. She prepares a sequencing reaction that includes the DNA template, primers, and DNA polymerase with chain-terminating nucleotides (dideoxynucleotides). As the reaction progresses, it generates fragments of different lengths, each ending with a fluorescently labeled terminator nucleotide.

5. **Capillary Electrophoresis**: To analyze the sequencing reaction products, Emma utilizes capillary electrophoresis. This technique separates the DNA fragments based on size as they pass through a narrow capillary filled with a polymer matrix. A laser excites the fluorescent terminators, and a detector records the emitted light, producing a sequencing readout.

6. **Bioinformatics Analysis**: Emma assembles and analyzes the sequence data using bioinformatics tools. She compares it to existing databases and identifies the gene responsible for the disorder. This discovery could potentially lead to breakthroughs in understanding and treating the rare disorder.

Conclusion

In this way, Emma successfully overcomes the challenge of identifying and sequencing the gene of interest using a combination of chromatography and sequencing techniques, contributing to the field of gene discovery.

Scenario-11

Meet Emily, a dedicated PhD researcher at a prestigious molecular biology laboratory. Her current quest revolves around **understanding the migration patterns of a newly discovered protein within biological samples**. The protein's localization is thought to play a crucial role in a rare genetic disorder.

Scientific challenge

Emily's supervisor hands her a vial containing precious protein samples and challenges her to decipher the migration behavior of this protein using electrophoresis, coupled with advanced microscopy techniques.

Scientific Approach

Emily embarks on her scientific journey with a well-designed plan. She starts by preparing her protein samples and running them through an SDS-PAGE gel. After electrophoresis, the separated protein bands are ready to be visualized.

1. **Gel Staining**: Emily carefully stains the gel using a protein-specific dye that binds to the separated proteins. This staining reveals distinct bands, each representing a different protein.

2. **Microscopy Exploration**: Emily doesn't stop at the gel itself. She uses a high-resolution fluorescence microscope to capture detailed images of the gel. By employing a specialized filter set, she ensures that the fluorescent signal from the protein-specific dye is captured accurately.

3. **Image Analysis**: Armed with a collection of gel images, Emily turns her attention to image analysis software. She processes the images, extracting quantitative data about the migration distance of each protein band. This data will be crucial in comparing the migration of the target protein with other known proteins.

4. **Localization Insight**: Emily's meticulous approach pays off as she uncovers the migration pattern of the target protein. Its distinctive position within the gel, when combined with her microscopy data, gives her insight into the protein's potential localization within the cell.

5. **Validation and Future Directions**: Emily's findings provide strong evidence for the protein's cellular localization. To validate her results, she designs additional experiments, including immunofluorescence assays, to visualize the protein directly within cells. The newfound knowledge sets the stage for further research into the genetic disorder and potential therapeutic strategies.

Conclusion

In this exciting journey, Emily's integration of electrophoresis with advanced microscopy techniques proves to be a powerful combination, shedding light on the intricate world of protein migration and localization.

Scenario-12

Megan, a dedicated Ph.D. candidate at a leading research institute, is engrossed in her study on **developing novel drug candidates to treat a rare autoimmune disorder**. Her experimental approach involves assessing how these drug candidates influence key cell signaling pathways in immune cells.

However, she's confronted with the challenge of accurately measuring the changes in signaling molecules upon drug exposure.

Scientific approach

Megan's quest begins by cultivating immune cells in a controlled environment using cell culture techniques. She opts for primary human T cells, as they play a pivotal role in the autoimmune disorder she's targeting. To address the challenge of evaluating signaling pathway changes, she designs a comprehensive experimental plan:

1. **Cell Culture Optimization**: Megan carefully maintains her T cells in a growth medium that mimics their natural environment. She ensures that the cells are healthy and responsive to external stimuli.

2. **Drug Exposure Protocol**: She exposes the cultured T cells to various concentrations of her drug candidates, as well as control samples. This step aims to replicate the conditions under which the drugs would act in the human body.

3. **Cell Lysis and Protein Extraction**: After the exposure period, Megan extracts proteins from the treated cells. She lyses the cells to release the intracellular contents, including signaling molecules involved in the interested pathways.

4. **Immunoprecipitation and Western Blotting**: Megan employs immunoprecipitation (IP) to isolate specific target proteins from the cell lysate. This allows her to focus on the key signaling molecules she's investigating. She then performs Western blotting to detect and quantify the levels of these proteins.

5. **Quantitative Analysis**: Using specialized software, Megan analyzes the Western blot results to quantify the changes in protein expression levels. She compares the treated samples with control samples to identify any significant alterations.

6. **Data Interpretation**: With the quantitative data in hand, Megan interprets the results in the context of the cell signaling pathways she's studying. She correlates changes in protein expression with potential impacts on the pathways' functionality.

Conclusion

By meticulously combining immunotechniques with cell culture, Megan overcomes the challenge of assessing the impact of her drug candidates on cell signaling pathways. Her thorough scientific approach provides valuable insights into the effectiveness of the drug candidates in modulating immune responses, paving the way for potential therapeutic breakthroughs in treating the autoimmune disorder.

Scenario-13

Meet Sarah, a dedicated Ph.D. researcher at the forefront of molecular biology. She's fascinated by the intricate dance between DNA and proteins within the cell, but one challenge has been keeping her up at night: **understanding the structural details of DNA-protein complexes**.

Scientific challenge

In her project, Sarah is investigating a crucial DNA-binding protein involved in DNA repair. However, the exact nature of the interactions between this protein and DNA remains elusive. Standard structural biology techniques like X-ray crystallography and NMR provide valuable information, but they often struggle with large and dynamic complexes like DNA-protein interactions.

Scientific approach

1. **UV-Visible Spectroscopy**: Sarah employs UV-Visible spectroscopy to monitor the changes in absorbance of the DNA-protein complex. By carefully selecting the appropriate wavelengths, she can observe the shifts in absorbance that occur when the protein binds to the DNA. These shifts can indicate changes in the local environment around the DNA bases, providing insights into the binding process.

2. **Fluorescence Spectroscopy**: To delve deeper, Sarah uses fluorescence spectroscopy. She labels the DNA and protein with fluorophores and quenchers, respectively. When the protein binds to the DNA, the fluorophore's emission is altered due to changes in its microenvironment. By measuring these changes, Sarah gains information about the distance and orientation between the two biomolecules.

3. **Molecular Dynamics Simulations**: To complement her experimental data, Sarah employs molecular dynamics simulations. By simulating the interactions between the DNA and protein, she can validate her spectroscopic findings and visualize the dynamic behavior of the complex over time. This helps her understand how the complex adapts to different conditions.

4. **Cross-Validation**: Sarah understands the importance of cross-validation. By comparing the insights gained from both spectroscopic methods and simulations, she can piece together a comprehensive picture of the DNA-protein

interactions. This holistic approach helps her validate her findings and draw reliable conclusions.

Conclusion

In the end, Sarah's multidisciplinary approach, combining advanced spectroscopic techniques and computational simulations, allows her to unravel the intricate world of DNA-protein interactions. Her work not only contributes to a deeper understanding of cellular processes but also serves as an inspiration for young scientists venturing into the realm of molecular biology.

Scenario-14

Meet Sarah, a dedicated PhD student at a renowned research institution. Her project involves **investigating the immune cell populations in patients with a rare autoimmune disorder**. To better understand the dynamics of immune cells, she needs to isolate and characterize specific immune cell subsets from blood samples.

Scientific challenge

Sarah's challenge is that the immune cell population she's interested in is rare and present in low quantities in the bloodstream. She needs a way to efficiently isolate these cells for further analysis. Her advisor suggests a combination with centrifugation and immunotechniques to overcome this hurdle.

Scientific approach

1. **Sample Preparation**: Sarah begins by carefully collecting blood samples from patients and healthy controls. She adds anticoagulants to prevent clotting and maintains a sterile environment to preserve cell integrity.

2. **Centrifugation**: She employs differential centrifugation, a technique that takes advantage of the different densities of blood components. By spinning the blood samples at varying speeds and durations, she separates the blood into distinct layers: plasma, buffy coat (white blood cells), and red blood cells.

3. **Buffy Coat Collection**: The buffy coat, containing the immune cells, is carefully extracted and transferred to a new tube. This step concentrates the immune cells, making further processing more effective.

4. **Immunotechniques** - Antibody Labeling: To specifically target the desired immune cell population, Sarah uses monoclonal antibodies labeled with fluorescent markers. These antibodies are designed to bind to surface markers specific to the target immune cells.

5. **Cell Sorting**: She employs flow cytometry, a powerful immunotechnology, to analyze and sort cells based on their fluorescent marker labeling. Flow cytometry can rapidly analyze thousands of cells per second, allowing for isolation of the rare immune cell population.

6. **Isolation and Culturing**: Once sorted, the isolated immune cells are collected and cultured in appropriate growth media. This step enables Sarah to expand the cell population for further characterization and functional studies.

Conclusion

By combining centrifugation to concentrate immune cells and immunotechniques to specifically identify and isolate the desired cell population, Sarah successfully tackles the challenge of isolating rare immune cells for immunophenotyping. This approach enables her to delve deeper into understanding the immune dynamics of the autoimmune disorder. Sarah's journey exemplifies how the synergy between traditional techniques like centrifugation and cutting-edge immunotechnologies can pave the way for groundbreaking research in immunology.

Scenario-15

Emily, a Ph.D. candidate at a prestigious research institution, is investigating the **subcellular localization of a novel protein involved in cell signaling**. The protein's role in various cellular compartments remains unclear, and her research aims **to correlate its expression levels with its specific cellular localization**. This is a critical step in understanding its function and potential implications in disease.

Scientific challenge

Emily's challenge is to design an experiment that not only quantifies the protein's expression levels but also precisely determines its location within different cellular compartments. She must choose the appropriate microscopy and blotting techniques to tackle this intricate problem.

Scientific approach

1. **Sample Preparation**: Emily starts by culturing cells under different conditions to induce changes in protein expression. She carefully selects cell lines and treatment protocols to ensure meaningful results.

2. **Microscopy for Localization**: Emily uses fluorescence microscopy to visualize the protein's distribution within the cells. She employs

immunofluorescence staining with specific antibodies targeting her protein interest. By choosing appropriate fluorophores and imaging settings, she can capture high-resolution images of the protein's localization.

3. **Quantification of Expression**: To quantify protein expression levels, Emily plans to perform a Western blot analysis. She extracts protein samples from different cell fractions, such as cytoplasmic, nuclear, and membrane fractions. After separation using gel electrophoresis, she transfers the proteins to a membrane and performs immunoblotting using the same antibodies as in her microscopy experiments.

4. **Data Integration**: Emily compares the fluorescence microscopy images with the Western blot results. By analyzing the intensity of fluorescence signals in different cellular compartments and correlating them with the corresponding bands on the blot, she can determine the relative expression levels of the protein in each compartment.

5. **Normalization and Statistical Analysis**: To ensure accuracy, Emily normalizes her data using appropriate internal controls and housekeeping proteins. She employs statistical tools to assess the significance of her findings and perform a robust correlation analysis.

6. **Cellular Dynamics**: By repeating the experiment over time and under various conditions, Emily gains insights into how the protein's expression and localization change dynamically in response to stimuli.

7. **Validation**: To validate her findings, Emily employs additional techniques such as confocal microscopy and alternative blotting methods. This comprehensive approach enhances the reliability of her results.

Conclusion

Emily's innovative approach of combining microscopy and blotting techniques allows her to tackle the complex challenge of correlating protein expression levels with cellular localization. Her meticulous experimental design and thoughtful analysis provide valuable insights into the protein's role within different cellular compartments, advancing our understanding of cell signaling pathways and potential therapeutic targets.

Scenario-16

Meet Sarah, a dedicated PhD student at a leading research institute, who is passionate about unraveling the mysteries of cellular behavior. Her current project involves **investigating how changes in growth conditions impact protein expression in a specific type of human cell**.

Scientific challenge

Sarah's challenge is to understand how alterations in growth conditions affect the expression of a vital protein involved in cellular signaling pathways. She needs to determine if certain conditions trigger changes in the protein's abundance, potentially shedding light on new avenues of research and therapeutic interventions.

Scientific approach

1. **Hypothesis Formulation**: Sarah begins by formulating a hypothesis based on existing knowledge. She hypothesizes that specific growth conditions will lead to noticeable changes in the expression of the target protein.

2. **Experimental Design**:
A. **Cell Culture Preparation**: Sarah cultures the human cells in different conditions, including varying nutrient concentrations, growth factors, and oxygen levels. She sets up control groups with standard conditions to compare the results.
B. **Protein Extraction**: After growing the cells, Sarah carefully extracts proteins from each sample using techniques that maintain protein integrity.
C. **Electrophoresis Setup**: For each sample, Sarah prepares gel matrices suitable for electrophoresis. She uses Agarose gel electrophoresis for an initial separation and then employs SDS-PAGE to achieve higher resolution separation.

3. **Electrophoresis Procedure**:
A. **Agarose Electrophoresis**: Sarah loads the extracted proteins onto the Agarose gel and applies an electric field. Since Agarose gel is effective for larger protein separations, this step helps her achieve an initial separation of the proteins.

B. **SDS-PAGE**: Next, she takes the separated proteins from the Agarose gel and further separates them using SDS-PAGE. This technique standardizes protein charge-to-mass ratios, allowing for a more accurate analysis of protein expression changes.

4. **Data Analysis**:
A. **Gel Imaging**: Sarah visualizes the separated protein bands using staining methods or specialized dyes. She captures images of the gels.

B. Band Intensity Measurement: Using image analysis software, Sarah quantifies the intensity of each protein band. Higher band intensity indicates greater protein abundance.

5. **Comparison and Interpretation**:
A. **Statistical Analysis**: Sarah compares the band intensities between different growth conditions and control groups. Statistical tests help her determine if observed differences are significant.
B. **Correlation to Growth Conditions**: By correlating the changes in protein expression with specific growth conditions, Sarah identifies conditions that induce substantial alterations in protein abundance.

Conclusion

Through the combination of electrophoresis techniques and careful cell culture experimentation, Sarah successfully uncovers how changes in growth conditions influence the expression of the target protein. Her findings contribute to the understanding of cellular responses and pave the way for potential therapeutic strategies targeting these proteins under different physiological contexts.

Scenario-17

Meet Sarah, a dedicated PhD student at a prestigious research institution. Her project revolves around developing a diagnostic assay for a rare autoimmune disease. To achieve this, she needs highly pure antibodies to serve as detection agents. However, **obtaining pure antibodies from complex biological samples** is a formidable challenge.

Scientific challenge

Sarah needs to purify antibodies from a mixture of serum proteins obtained from patients with the autoimmune disease. These antibodies are crucial for the specificity of her diagnostic assay. The challenge lies in achieving high purity while preserving the antibodies' structural and functional integrity.

Scientific approach

Sarah designs a comprehensive approach that combines Chromatography and Immunotechniques to tackle this challenge effectively.
1. **Initial Sample Preparation**: Sarah starts by collecting blood serum samples from patients. She centrifuges the samples to remove cellular components, obtaining a serum containing a mixture of proteins, including the interested antibodies.

2. **Affinity Chromatography**: Sarah employs affinity chromatography as the primary purification step. She immobilizes antigen molecules (related to the autoimmune disease) onto a solid matrix, creating an affinity column. As the serum passes through the column, antibodies with affinity for the antigen bind to it while other proteins flow through. By washing the column, she removes non-specifically bound proteins.

3. **Elution of Antibodies**: To elute the antibodies specifically, Sarah uses a mild elution buffer that disrupts the antigen-antibody interactions without denaturing the antibodies. This yields a solution enriched with the desired antibodies.

4. **Ion Exchange Chromatography**: Sarah performs a subsequent ion exchange chromatography step to further purify the eluted antibodies. This technique separates molecules based on their charge properties. By adjusting the pH and ionic strength of the buffer, she controls the binding and elution of the antibodies, achieving higher purity.

5. **Immunotechniques Validation**: Sarah validates the purified antibodies' specificity and functionality using techniques such as Enzyme-Linked Immunosorbent Assay (ELISA) and Western Blot. This ensures that the antibodies retained their binding capability and were not denatured during purification.

Conclusion

Sarah successfully purifies antibodies with high specificity and purity from complex serum samples. These purified antibodies are then integrated into her diagnostic assay, which exhibits remarkable sensitivity and accuracy in detecting the autoimmune disease biomarkers. By combining Chromatography and Immunotechniques in her scientific approach, Sarah overcomes the challenge to obtaining pure antibodies for her diagnostic assay, making significant strides in the field of medical diagnostics.

Scenario-18

Emily, a dedicated PhD student at a prestigious research institute, is fascinated by the complex world of RNA molecules and their secondary structures. She is faced with a challenging task: **to uncover the intricate secondary structures of a newly discovered RNA molecule** that plays a crucial role in a disease pathway. Emily's goal is **to understand how these structures influence the RNA's function and potential therapeutic applications**.

Scientific approach

To tackle this scientific challenge, Emily devises a comprehensive approach that combines Spectroscopy and Sequencing Techniques:

1. **UV Absorbance Spectroscopy**: Emily starts by using UV absorbance spectroscopy to analyze the absorption of UV light by the RNA molecule. This allows her to deduce information about the RNA's nucleotide composition and its overall structural characteristics.

2. **Circular Dichroism (CD) Spectroscopy**: Emily then employs CD spectroscopy to gain insights into the secondary structures of the RNA molecule. CD spectroscopy measures the differential absorption of left- and right-circularly polarized light by chiral molecules like RNA. By analyzing the CD spectra, she can determine the RNA's secondary structure elements, such as helices and loops.

3. **Fluorescence Spectroscopy**: Emily integrates fluorescence spectroscopy to probe the microenvironment around specific nucleotides within the RNA. By using fluorescent labels at strategic positions, she can monitor changes in fluorescence intensity, wavelength, or lifetime, which reveal conformational changes in the RNA as it folds and interacts with other molecules.

4. **High-Throughput Sequencing**: Emily utilizes high-throughput sequencing techniques to complement the spectroscopic data. By sequencing a large number of RNA molecules, she can identify common structural motifs and assess their prevalence within the RNA population. This provides a statistical perspective on the secondary structure ensemble.

Conclusion

Emily combines the information gathered from the spectroscopic techniques and sequencing data to construct a 3D model of the RNA's secondary structure. The UV absorbance and CD spectroscopy help identify major structural features, while fluorescence spectroscopy provides insights into local conformational changes. The high-throughput sequencing data confirm the prevalence of certain secondary structure motifs.

Scenario-19

Meet Sarah, a dedicated PhD student at a leading research institute. Her project involves **understanding the molecular basis of a rare genetic disorder that leads to dysregulated gene expression in certain cells**. Sarah suspects that the abnormal gene expression might be linked to changes in RNA stability and processing.

Scientific approach

1. **Sample Preparation and Centrifugation**: Sarah begins by culturing cells from both healthy individuals and those affected by the genetic disorder. She knows that to isolate intact RNA, she needs to ensure the cells are handled carefully. After harvesting the cells, she uses a centrifuge to separate cellular components based on their density. Centrifugation aids in isolating the nuclei, where the genetic information resides, from other cellular components.

2. **RNA Extraction**: Once the nuclei are isolated, Sarah employs a molecular technique known as "RNA extraction" to collect RNA molecules. This is where she faces her first challenge. The dysregulated gene expression leads to lower RNA stability, making it prone to degradation. To counteract this, she opts for a robust extraction method that includes the use of denaturing agents and high-quality reagents. These precautions help maintain RNA integrity during the isolation process.

3. **Reverse Transcription**: Sarah now has a collection of RNA, but she's not done yet. To analyze gene expression, she needs to convert the RNA into complementary DNA (cDNA) through a process called reverse transcription. This step allows her to work with DNA, which is more stable and suitable for various molecular analyses.

4. **Quantitative PCR (qPCR)**: With cDNA in hand, Sarah decides to employ quantitative PCR (qPCR), a powerful molecular technique. She designs specific primers to amplify the target genes of interest. Here's where her second challenge arises. The genes she's studying are present in low amounts due to the dysregulation. To tackle this, she optimizes the qPCR conditions, using stringent controls and carefully selecting housekeeping genes for normalization.

5. **Data Analysis and Interpretation**: After running the qPCR experiments, Sarah faces the final challenge: interpreting the data. The gene expression levels are subtle, and statistical analysis is crucial to distinguish significant changes from noise. Sarah employs advanced bioinformatics tools to analyze the large dataset generated by the qPCR experiments. She identifies patterns of dysregulated gene expression in the affected cells, providing insight into the genetic disorder's underlying mechanisms.

Conclusion

In this scenario, Sarah's scientific approach involves a combination with centrifugation and molecular techniques to isolate intact RNA from cells with dysregulated gene expression. Her careful experimental design, use of robust protocols, and bioinformatics analysis allow her to unveil crucial insights into the genetic disorder's molecular basis.

Scenario-20

Meet Alice, a passionate Ph.D. candidate at a cutting-edge research institute. She is determined **to decipher the intricate relationship between gene expression and cellular morphology** using advanced techniques.

Scientific challenges

Alice's challenge is to understand how specific genes are expressed within different cellular structures in a tissue sample. Traditional methods, like bulk RNA sequencing, provide information about overall gene expression but lack spatial context. Alice knows that to truly comprehend complex biological processes; she needs to develop an approach that links gene expression with the spatial arrangement of cells.

Scientific approach

Incorporating microscopy and sequencing techniques, Alice devises a strategic plan:

1. **Sample Preparation**: Alice starts by carefully obtaining tissue samples and fixing them in a way that preserves cellular morphology. She then prepares thin tissue sections for analysis.

2. **Microscopy Imaging**: Alice uses advanced microscopy techniques, such as confocal or multiphoton microscopy, to capture high-resolution images of the tissue sections. This step provides detailed information about cellular structures and their spatial distribution.

3. **Spatially Resolved Sequencing**: To correlate gene expression with cellular morphology, Alice leverages spatial transcriptomics. This technique allows her to analyze the gene expression of individual cells while preserving their spatial coordinates on the tissue section.

4. **Data Integration**: Alice collects massive amounts of data from microscopy and spatial transcriptomics. She develops sophisticated computational algorithms to integrate the data, aligning gene expression profiles with cellular locations.

5. **Correlation Analysis**: With integrated data in hand, Alice employs advanced statistical techniques to correlate gene expression patterns with cellular

morphologies. This step reveals insights into how gene activity influences cellular structure and function.

Conclusion

Through her meticulous work, Alice successfully uncovers a previously hidden relationship between gene expression and cellular morphology. She identifies key genes that are selectively expressed in specific cellular structures, shedding light on the molecular mechanisms that underlie tissue complexity and function. By merging microscopy and sequencing techniques, Alice's groundbreaking research paves the way for a new era of understanding in the field of spatial transcriptomics. Her innovative approach will inspire further investigations into the intricate interplay between genes and cellular architecture.

Scenario-21

Emily is a Ph.D. student at a renowned research institute, and her project involves studying a specific gene's expression pattern in different tissues of a newly discovered plant species. She needs **to verify the presence and quantity of a particular DNA fragment in the plant's genome**. This DNA fragment is suspected to be associated with the plant's unique adaptation to its environment. Emily's goal is to perform Southern blotting to confirm the presence of the DNA fragment in the genomic DNA extracted from various plant tissues.

Scientific approach

Emily faces the challenge of detecting a specific DNA fragment within a complex mixture of genomic DNA. Here's her scientific approach:

1. **Genomic DNA Extraction**: Emily starts by isolating genomic DNA from the plant tissues using standard extraction techniques. She ensures that the DNA remains intact and undegraded during the extraction process.

2. **DNA Digestion**: She performs restriction enzyme digestion on the extracted genomic DNA. This enzymatic treatment cuts the DNA at specific recognition sites, yielding fragments of varying lengths.

3. **Agarose Gel Electrophoresis**: Emily loads the digested DNA samples onto an agarose gel for electrophoresis. Applying an electric field causes the DNA fragments to migrate through the gel matrix, separating them based on size. Shorter fragments move farther from the loading well, while longer ones stay closer.

4. **Transfer to Membrane**: After electrophoresis, Emily uses the Southern blotting technique to transfer the separated DNA fragments from the gel to a nitrocellulose or nylon membrane. This immobilizes the DNA fragments in their relative positions on the membrane.

5. **Probe Design and Labeling**: Emily designs a DNA probe complementary to the target DNA fragment. The probe is labeled with a radioactive or fluorescent tag, allowing it to bind specifically to the complementary DNA sequence.

6. **Hybridization**: Emily incubates the labeled DNA probe with the membrane, allowing it to hybridize with any complementary DNA fragments immobilized on the membrane.

7. **Washing and Detection**: After hybridization, Emily washes away any unbound probe, leaving only the probe that has hybridized to the target DNA fragment. She then detects the presence of the labeled probe, which indicates the presence of the target DNA fragment in the plant's genome.

Conclusion

By combining electrophoresis with Southern blotting, Emily successfully confirms the presence of the specific DNA fragment in the plant's genomic DNA. This information contributes to a deeper understanding of the plant's genetic makeup and its adaptive traits.

Scenario-22

Meet Sarah, a dedicated PhD student working in a bustling lab at a renowned research university. She's passionately investigating the intricacies of gene regulation and wants **to unravel how certain proteins interact with DNA to control gene expression**. Her focus is on a specific transcription factor that plays a pivotal role in cell differentiation.

Scientific challenge

Sarah needs to determine whether the transcription factor of interest directly binds to a specific gene's promoter region in living cells. This involves identifying not only the gene sequence where the transcription factor binds but also understanding the functional consequences of this interaction on gene expression.

Scientific approach

To address this challenge, Sarah designs a comprehensive experimental plan that involves chromatin immunoprecipitation (ChIP), a powerful technique that combines immunology and molecular biology.

1. **Sample Preparation**: Sarah starts by growing and treating her cells to stabilize the protein-DNA interactions. She carefully cross-links the proteins to the DNA using formaldehyde.

2. **Cell Lysis and Chromatin Shearing**: She lyses the cells and fragments the chromatin into manageable sizes. This is a critical step to ensure that she captures the specific protein-DNA complexes she's interested in.

3. **Immunoprecipitation**: Sarah then introduces an antibody that specifically recognizes her transcription factor of interest. This antibody will pull down the protein-DNA complexes of interest while leaving other cellular components behind.

4. **DNA Purification**: After immunoprecipitation, Sarah reverses the cross-links, separating the protein from the DNA. This leaves with a mixture of DNA fragments enriched with the sequences where her transcription factor was bound.

5. **PCR or Sequencing**: Now, Sarah can analyze these DNA fragments in a couple of ways. She can perform PCR using primers designed for the promoter region of her gene of interest. Alternatively, she can sequence the DNA fragments using next-generation sequencing techniques to get a genome-wide view of where the transcription factor binds.

Conclusion

Sarah's meticulous approach using ChIP allows her to map binding sites of her transcription factor across the genome. By comparing the binding sites with gene expression data, she can infer whether the transcription factor's presence enhances or represses gene expression. This information is crucial for understanding how this protein-DNA interaction contributes to cellular differentiation processes. Through Sarah's innovative use of chromatin immunoprecipitation, she has taken a significant step toward deciphering the complex regulatory networks that govern gene expression, paving the way for future breakthroughs in the field of cell biology.

Scenario-23

Meet Maya, a dedicated PhD candidate at the Institute of Biotechnology. She's fascinated by the intricate world of cellular metabolism and is determined to uncover the mysteries of how cells respond to various environmental stimuli. Maya's current project involves **investigating metabolic changes in cultured cells using spectroscopic methods**.

Scientific challenge

Maya is confronted with a challenging problem: She aims to understand how different nutrient conditions influence the metabolic activity of cells. Specifically, she wants to assess the metabolic shift that occurs in cells when they're exposed to varying glucose concentrations. This information could provide crucial insights into disease mechanisms and potentially lead to innovative therapeutic approaches.

Scientific approach

Maya designs a meticulous experimental plan to tackle this challenge:
1. **Cell Culture Setup**: Maya cultures a line of human liver cells known for their metabolic versatility. She divides the cells into two groups: one group in a standard glucose medium and the other in a low glucose medium.

2. **Spectroscopic Analysis**: To monitor metabolic changes, Maya employs a combination of fluorescence spectroscopy and NMR spectroscopy. The former helps her assess the activity of key metabolic enzymes, while the latter provides a detailed metabolic fingerprint of the cells.

3. **Data Collection and Analysis**: For several days, Maya periodically samples cells from both groups. She analyzes the fluorescence emission patterns and NMR spectra to detect changes in enzyme activities and metabolite concentrations.

4. **Fluorescence Insights**: Maya notices that cells in the low glucose medium show decreased fluorescence intensity compared to the standard glucose medium. This suggests a downregulation in certain metabolic pathways, indicating the cells are conserving energy under nutrient stress.

5. **NMR Revelations**: Analyzing the NMR spectra, Maya identifies altered peaks corresponding to metabolites involved in glycolysis and the citric acid cycle. This confirms the cells' metabolic adaptation to the low-glucose environment.

6. **Integration of Data**: By integrating the fluorescence and NMR data, Maya creates a comprehensive metabolic profile. She maps out the changes in enzyme activities and metabolite concentrations, highlighting the cellular response to different glucose conditions.

Conclusion

Maya's meticulous approach not only sheds light on the cellular responses to varying glucose levels but also showcases the power of combining spectroscopy with cell culture. Her findings contribute to the understanding of metabolic flexibility and could potentially have implications for diabetes research and metabolic disorders.

Scenario-24

Sarah, a Ph.D. candidate at a prestigious research institute, is dedicated to improving protein production in cell cultures for medical applications. Her current project involves **optimizing media components to enhance protein yield**. She's specifically focused on using a combination of chromatography techniques and cell culture methods.

Scientific challenge

Sarah's challenge is to identify the ideal combination of chromatography techniques and cell culture conditions that will maximize the production of a valuable protein. She needs to figure out the most efficient chromatography approach and the optimal cell culture medium to achieve the highest yield while maintaining protein quality.

Scientific approach

1. **Preparation Phase**: Sarah first selects a variety of chromatography techniques, such as ion exchange chromatography and size exclusion chromatography, which can help her purify and separate the target protein from the cell culture medium.

2. **Experimental Design**: Sarah designs a comprehensive experiment that involves multiple conditions. She prepares different cell culture media by varying components like amino acids, vitamins, and growth factors. Each medium is then used to culture cells expressing the protein of interest.

3. **Cell Culturing**: Sarah cultures the cells in bioreactors, carefully controlling parameters like temperature, pH, and oxygen levels. She monitors cell growth and protein production over time.

4. **Harvesting and Purification**: Once cells have reached the desired growth phase, Sarah harvests the culture medium. She applies chromatography techniques to purify the protein. Different chromatography columns are tested with varying conditions to find the optimal separation parameters.

5. **Protein Analysis**: Sarah analyzes the purified protein samples using techniques like SDS PAGE to determine the protein's purity and concentration. She also assesses the protein's biological activity.

6. **Data Analysis**: Sarah compiles and analyzes the data collected from her experiments. She identifies the media components and chromatography techniques that resulted in the highest protein yield and quality.

7. **Optimization**: Based on the data analysis, Sarah identifies the most effective combination of chromatography techniques and cell culture media components. She optimizes the process by refining the parameters that lead to the best results.

Conclusion

In her report, Sarah outlines the optimized approach, detailing the specific chromatography methods, cell culture conditions, and media components that yielded the highest protein production. She discusses the scientific significance of her findings and their potential applications in the field of protein production and biotechnology. By successfully combining chromatography and cell culture, Sarah overcomes the challenge of enhancing protein production, contributing valuable insights to the scientific community's understanding of optimizing bioprocesses for medical and industrial purposes.

Scenario-25

Emily, a dedicated PhD student at a renowned research institute, is working on a groundbreaking project in the field of single-cell genomics. Her research goal is **to understand the complex gene expression patterns within heterogeneous cell populations at the single-cell level**. She's particularly interested in a specific cell population that plays a crucial role in a rare neurological disorder.

Scientific challenge

Isolating this specific cell type has proven to be a daunting task due to its scarcity in the tissue samples. Traditional methods of cell sorting have failed to yield a pure and viable population. Emily's challenge is to devise a method that not only separates these cells efficiently but also preserves their RNA integrity for subsequent single-cell RNA sequencing (scRNA-seq).

Scientific approach

Emily decides to combine the power of centrifugation and sequencing techniques to overcome this challenge. She carefully designs a multi-step protocol:

1. **Tissue Preparation**: Emily starts by carefully dissecting the target tissue and dissociating it into a single-cell suspension.

2. **Density Gradient Centrifugation**: She employs density gradient centrifugation to enrich the specific cell population. By layering the single-cell suspension onto a density gradient medium and centrifuging it, she's able to separate cells based on their buoyant density. The specific cell type of interest settles in a distinct layer.

3. **Cell Viability and RNA Integrity Assessment**: Emily employs fluorescence staining to confirm the viability and integrity of the enriched cells. She uses a combination of viability dyes and RNA-specific stains to ensure that the sorted cells are not only viable but also have preserved RNA.

4. **Single-Cell RNA Sequencing (scRNA-seq)**: Emily performs scRNA-seq on the enriched cell population. She uses a state-of-the-art platform that can process individual cells in parallel, generating comprehensive gene expression profiles.

5. **Data Analysis**: Emily processes the scRNA-seq data, using advanced bioinformatics tools to identify unique gene expression patterns within the enriched cell population. This analysis helps her uncover crucial insights into the molecular characteristics of the specific cell type.

Conclusion

By combining centrifugation for cell enrichment and scRNA-seq for molecular profiling, Emily successfully overcomes the challenge of isolating a rare cell population and gains a deeper understanding of its role in neurological disorders. Her innovative approach sets a new standard for studying rare cell populations in complex tissues, offering potential breakthroughs in the field of neurobiology.

Scenario-26

Sarah is a dedicated PhD student at a prestigious research institution. Her project focuses on **unraveling the mysteries of cellular organelles at the nanoscale level**. She's particularly interested in observing the **intricate interactions between mitochondria and the endoplasmic reticulum (ER) within living**

cells. Traditional microscopy techniques have limitations in providing the necessary resolution to see these tiny structures in detail.

Scientific challenge

Sarah's challenge is to find a way to visualize the interactions between mitochondria and the ER in living cells with unprecedented detail, down to the nanoscale level. She needs to overcome the diffraction limit of conventional microscopy methods that prevent her from distinguishing structures that are very close together.

Scientific Approach

To tackle this challenge, Sarah devises a comprehensive scientific approach:
1. **Super-Resolution Microscopy Selection**: Sarah chooses a suitable super-resolution microscopy technique, such as STED (Stimulated Emission Depletion) microscopy or SIM (Structured Illumination Microscopy), known for their ability to surpass the diffraction limit and provide images with enhanced resolution.

2. **Sample Preparation**: Sarah carefully prepares her samples, including live cells expressing fluorescently tagged mitochondrial and ER proteins. She ensures the cells are healthy and well-maintained for accurate observations.

3. **Labeling and Imaging**: Using molecular techniques, Sarah introduces the appropriate fluorescent markers to label the mitochondria and ER structures. She optimizes the labeling process to achieve high specificity and minimal interference with cellular function. Then, she performs imaging using the selected super-resolution microscopy technique.

4. **Data Acquisition and Analysis**: She captures high-resolution images of the cellular structures. These images reveal intricate details of the interactions between mitochondria and the ER that were previously unobservable. She uses advanced image analysis software to quantify the proximity, frequency, and nature of these interactions.

5. **Validation and Reproducibility**: To ensure the accuracy and reproducibility of her findings, Sarah repeats the experiment multiple times, analyzing different cells and regions. She also collaborates with colleagues to verify her results and interpretations.

Conclusion

With her thorough approach, Sarah successfully uncovers new insights into the dynamic interactions between mitochondria and the ER at the nanoscale level.

She compiles her findings, along with the methodologies used, into a comprehensive research paper, which she submits to a respected scientific journal. In this way, Sarah combines super-resolution microscopy with molecular techniques to overcome the challenges of visualizing cellular structures at the nanoscale, ultimately contributing to the advancement of our understanding of cell biology.

Scenario-27

Meet Sarah, a dedicated PhD student at a cutting-edge research institute. She's passionate about contributing to the field of medical diagnostics. One day, her mentor, Dr. Anderson, presents her with a challenging research problem. There's a pressing need **to develop a method for detecting specific antigens in patient blood samples, which could provide early indications of a complex autoimmune disorder that currently lacks efficient diagnosis**.

Scientific challenge

Sarah is intrigued by the challenge and realizes that combining Electrophoresis and Immunotechniques might hold the key. The autoimmune disorder is characterized by the presence of certain antigens that are usually absent in healthy individuals. However, these antigens are present in trace amounts and are difficult to detect.

Scientific approach

1. **Sample Preparation**: Sarah starts by carefully processing patient blood samples to extract relevant components. She separates the serum, where antigens are likely to be present.

2. **Electrophoresis**: She utilizes electrophoresis to separate the serum proteins based on their charge and size. This initial separation provides a way to narrow down the potential regions where the target antigens might be present.

3. **Transfer to Membrane**: After electrophoresis, Sarah transfers the separated proteins onto a membrane. This allows her to work with the proteins in a more controlled manner.

4. **Immunoblotting**: Now comes the crucial step. Sarah employs immunoblotting techniques to introduce specific antibodies that can bind to the antigens of interest. By using different antibodies with known specificities, she can detect a range of potential antigens associated with the autoimmune disorder.

5. **Signal Detection**: To visualize the antigens, Sarah employs chemiluminescence. This technique creates a light signal where the antibodies have bound to antigens on the membrane. By analyzing the pattern of signals, she can identify the presence of specific antigens associated with the disorder.

6. **Quantification**: Sarah develops a quantification method based on the intensity of the signals. This allows to detect the presence of antigens and to assess their levels, which could potentially correlate with disease severity.

Conclusion

Sarah's innovative approach proves successful. She identifies a specific antigen signature that is consistently present in patients with the autoimmune disorder but absent in healthy individuals. This breakthrough discovery paves the way for a new diagnostic test that provides early indications of the disorder, enabling timely interventions and improved patient outcomes.

Scenario-28

Meet Sarah, a dedicated PhD student at a prestigious research institute. She's passionate **about unraveling the mysteries of glycosylation in biological samples**. Glycosylation plays a crucial role in protein function and cellular communication, but it's a complex process with various structural nuances that can impact protein behavior.

Scientific challenge

Sarah's supervisor challenges her to investigate the glycosylation patterns of a specific protein linked to a rare genetic disorder. The disorder's underlying mechanism remains poorly understood, and glycosylation irregularities might be at the heart of it.

Scientific approach

1. **Sample Preparation**: Sarah starts by isolating protein samples from both healthy and affected tissues. She employs chromatography techniques to separate the glycosylated proteins based on their charge, size, and hydrophobicity. High-performance liquid chromatography (HPLC) with a specialized glycoprotein column helps her achieve the initial separation.

2. **Identification and Isolation**: After separation, Sarah uses mass spectrometry to identify glycosylated protein peaks. This provides insights into the protein's mass and glycan composition. She's able to pinpoint potential differences between healthy and affected samples.

3. **Blotting Techniques**: Next, Sarah employs western blotting and lectin blotting to assess glycosylation patterns. Western blotting reveals the specific protein's presence and gives a rough idea of its molecular weight. Lectin blotting, on the other hand, employs lectin proteins that bind to specific glycan structures. By comparing the lectin binding patterns between healthy and affected samples, Sarah can infer glycosylation changes associated with the disorder.

4. **Data Integration**: Sarah integrates the chromatography and blotting data, forming a comprehensive picture of the glycosylation differences. This helps her identify specific glycosylation variants associated with the disorder and potential therapeutic targets.

5. **Validation**: To validate her findings, Sarah designs experiments to modulate glycosylation in cell cultures. She uses chromatography and blotting techniques to confirm that altering glycosylation patterns affects the protein's behavior and interactions.

Conclusion

Through her innovative combination of chromatography and blotting techniques, Sarah uncovers crucial insights into the role of glycosylation in the rare genetic disorder. Her work contributes to the understanding of the disorder's molecular mechanisms and opens new avenues for therapeutic interventions. By skillfully merging chromatography and blotting techniques, Sarah not only addresses her scientific challenge but also advances the field's understanding of glycosylation's impact on protein function and human health.

Scenario-29

Emily is a Ph.D. candidate at a prestigious research institute, focused on unraveling the mysteries of genetic material. She's been working diligently on understanding how different genes are expressed under various conditions. One day, her advisor hands her a new challenge: **quantify the expression levels in a specific gene under two different experimental treatments**.

Scientific challenge

The catch is that the gene expression levels are expected to be relatively low, and traditional methods might not be sensitive enough to detect subtle changes. Emily knows that UV-absorption spectroscopy can provide quantitative information about the concentration of nucleic acids. However, the low concentration of the target gene's product poses a significant challenge.

Scientific approach

Emily formulated a comprehensive plan to tackle this challenging experiment.
1. **Sample Preparation**: She starts by carefully extracting nucleic acids from the samples representing the two experimental conditions. She uses a reliable method that ensures the integrity of the nucleic acids during extraction.

2. **Denaturation and Blotting**: Emily performs gel electrophoresis to separate the nucleic acids based on size. Once separated, she employs Southern blotting to transfer the nucleic acids onto a membrane. This immobilizes the nucleic acids in a way that preserves their relative positions on the gel.

3. **Hybridization and Detection**: To specifically detect the target gene's nucleic acid, Emily uses a labeled probe that complements the gene's sequence. The probe is designed to bind only to the target sequence. After hybridization, she washes away any unbound probe.

4. **UV-Absorption Spectroscopy**: Now that Emily has the labeled target gene immobilized on the membrane, she carefully places the membrane into a UV-absorption spectroscopy instrument. The UV light will cause the labeled nucleic acids to emit fluorescence, allowing her to quantify their concentration.

5. **Quantitative Analysis**: Emily compares the fluorescence intensity of the labeled nucleic acids from the two experimental conditions. By correlating the intensity with the known concentrations of a control sample, she can determine the concentration of the target gene's nucleic acids for each experimental condition.

6. **Data Interpretation**: Using the concentration data, Emily can now infer the differences in gene expression levels between the two experimental treatments. She conducts statistical analyses to validate the significance of her findings.

Conclusion

By combining the power of UV-absorption spectroscopy with blotting techniques, Emily successfully overcomes the challenge of low nucleic acid concentrations. Her innovative approach provides valuable insights into the gene expression dynamics under different experimental conditions.

Scenario-30

Alice, a dedicated PhD student at a prestigious research institution, faces an exciting yet complex challenge. She is determined to unravel the mysteries of a cell's inner workings by analyzing its organelles' proteomes. Her goal is **to**

understand how protein expression varies across different cellular compartments. To achieve this, she must overcome the daunting task to isolate and analyze these organelles with precision.

Scientific approach

Alice decides to tackle the challenge by combining the power of centrifugation and molecular techniques. Her plan involves fractionating the cell's components into distinct organelles, followed by a deep dive into their proteomes. Here's her proposed approach:

1. **Sample Preparation**: Alice starts by preparing a cell lysate and gently homogenizing it to break open the cells while preserving integrity of organelles.

2. **Differential Centrifugation**: She employs differential centrifugation, a technique that exploits differences in organelle density and size to separate them. After a series of carefully timed and controlled centrifugation steps, Alice obtains a series of pellets enriched with specific organelles.

3. **Isolation of Organelles**: Alice meticulously resuspends each pellet and further purifies organelles using density gradient centrifugation. This technique helps her achieve better separation of organelles that might have similar densities.

4. **Protein Extraction**: With purified organelles in hand, Alice employs various molecular techniques to extract proteins from each fraction. She carefully maintains the native conditions of the proteins to ensure accurate representation.

5. **Proteomic Analysis**: Alice's next step involves using advanced mass spectrometry techniques. She identifies and quantifies the proteins present in each organelle fraction. This data-rich analysis provides insights into the unique protein composition of each organelle.

6. **Data Integration**: Armed with the proteomic data, Alice constructs a comprehensive map of protein distribution across different organelles. She identifies organelle-specific proteins and uncovers potential interactions and pathways that shape cellular functions.

Conclusion

By skillfully combining centrifugation and molecular techniques, Alice successfully overcomes the experimental challenge of organelle proteome analysis. Her dedication and innovative approach contribute to the growing understanding of cellular complexity and pave the way for future breakthroughs in biology and medicine.

Scenario-31

Sophia, a dedicated Ph.D. researcher at a prestigious biomedical institute, is determined to unravel the intricate interactions of immune cells within tumor microenvironments. Her project focuses on understanding the dynamic changes in immune cell populations using cutting-edge techniques. Her supervisor presents her with a challenge: **to simultaneously visualize and profile multiple immune cell types within tumor samples to gain a comprehensive view of their spatial distribution and activation states**.

Scientific challenge

Sophia is facing a complex challenge: How can she accurately characterize the diverse immune cell populations within tumor tissues and gain insights into their roles using microscopy and immunotechniques! Propose a scientific approach that enables her to perform multiplexed immunofluorescence, combining high-resolution imaging with advanced immunostaining methods.

Scientific approach

To address this challenge, Sophia can employ a strategic scientific approach that synergizes microscopy and immunotechniques:

1. **Sample Preparation**: Sophia will carefully prepare tumor tissue sections for analysis. These sections will be treated to ensure optimal antigen retrieval, preserving cellular morphology and antigenicity.

2. **Antibody Selection**: Sophia will choose a panel of well-validated antibodies, each labeled with distinct fluorophores. These antibodies will target specific immune cell markers to differentiate between various populations, enabling multiplexed staining.

3. **Sequential Staining and Imaging**: Sophia will perform sequential rounds of immunostaining, carefully removing or quenching fluorescence after each staining cycle to prevent signal overlap. Between each staining step, she will image the samples using high-resolution microscopy.

4. **Image Analysis**: Sophia will employ sophisticated image analysis software to deconvolute the multiplexed signals. She will use colocalization algorithms to precisely identify different immune cell populations and assess their spatial relationships.

5. **Quantitative Profiling**: Through quantification of signal intensities, Sophia will obtain data on the abundance and activation status of immune cells across the tumor tissue. This will allow her to map out the immune landscape in unprecedented detail.

6. **Data Integration**: Sophia will integrate her multiplexed immunofluorescence data with other omics data, such as gene expression profiles, to gain a holistic understanding of the immune response within the tumor microenvironment.

7. **Hypothesis Testing**: Using the spatial and activation information gained from her analysis, Sophia can formulate hypotheses about immune cell interactions, response patterns, and potential therapeutic targets.

Conclusion

By strategically combining advanced microscopy with multiplexed immunofluorescence techniques, Sophia can overcome the challenge to characterizing immune cell populations within tumor tissues. Her approach will provide insights into the complex interplay between different immune cells and their contributions to tumor immunity.

Scenario-32

Emily, a dedicated PhD student at a prestigious research institute, is tasked with solving a perplexing mystery. The institute has been receiving reports on **genetically identical crops showing up in different fields, leading to concerns about potential cross-contamination**. Emily's supervisor believes that **DNA fingerprinting** could shed light on this situation.

Scientific challenge

Emily's challenge is to determine whether the genetically identical crops are indeed the same, and if so, trace their origin to identify the source for contamination.

Scientific approach

1. **PCR Amplification**: Emily starts by isolating DNA from the crops in question. She then performs Polymerase Chain Reaction (PCR) to amplify specific regions of the DNA that are unique to each crop. This will generate a collection of DNA fragments that she can analyze.

2. **Electrophoresis**: Emily runs the PCR products through Agarose Gel Electrophoresis. This technique separates DNA fragments based on their size.

By comparing the fragment sizes of the genetically identical crops, she can determine if they have identical DNA profiles.

3. **DNA Sequencing**: To get a more detailed understanding, Emily chooses a subset of PCR products and sends them for DNA sequencing. She uses a next-generation sequencing platform to obtain the entire DNA sequence of these fragments.

4. **Sequence Comparison**: Emily aligns the DNA sequences from the genetically identical crops with known reference sequences. This allows her to identify any variations or mutations that might have occurred. She also uses bioinformatics tools to compare the sequences and identify any differences.

5. **Origin Tracing**: By comparing the DNA sequences, Emily can trace the origin of the contamination. If she finds identical sequences between the crops and a specific reference source, it indicates that the contamination likely originated from that source.

6. **Reporting and Validation**: Emily compiles her findings and prepares a comprehensive report for her supervisor. To validate her results, she repeats the entire process with different samples and cross-validates the data to ensure accuracy.

Conclusion

Through a combination of electrophoresis and DNA sequencing techniques, Emily successfully unravels the mystery behind the genetically identical crops. Her research not only identifies the source of contamination but also establishes a robust methodology for DNA fingerprinting in agricultural settings.

Scenario-33

Sophia, a dedicated Ph.D. researcher at a prestigious research institute, was intrigued by the prospect of combining chromatography and microscopy to solve a complex puzzle in her field of study. She had been studying the dynamics of cellular transport for years, and her current challenge revolved around **identifying specific compounds within a mixture that played a crucial role in intracellular signaling**.

Scientific challenge

The mixture contained fluorescently labeled compounds that were notoriously difficult to separate due to their similar properties. Sophia's goal was to isolate these compounds and gain a deeper understanding of their behavior within living cells. This was crucial for unlocking insights into a range of diseases, including those related to cellular communication.

Scientific approach

Sophia approached the challenge with a scientific strategy that integrated chromatography and microscopy, leveraging the strengths of both techniques. Her approach was as follows:

1. **Preparation and Chromatography**: Sophia first utilized a specialized chromatographic technique that was renowned for its high-resolution separation capabilities. This technique involved a stationary phase with specific interactions suitable for separating fluorescently labeled compounds. She carefully optimized the mobile phase composition and flow rate to achieve maximum separation efficiency.

2. **Fraction Collection and Analysis**: As the chromatography process proceeded, Sophia collected fractions that contained the separated compounds. Each fraction represented a unique mixture of compounds. To confirm the presence of the fluorescently labeled compounds in each fraction, she employed spectroscopic techniques such as UV-visible and fluorescence spectroscopy. This step allowed her to assess the distribution of her compounds of interest throughout the chromatographic run.

3. **Microscopic Imaging**: After identifying the fractions enriched with her compounds of interest, Sophia turned to advanced microscopy techniques. She used confocal microscopy to visualize the distribution of the labeled compounds within living cells. By tracking the movement of the compounds in real time, she gained insights into their roles in intracellular signaling pathways.

4. **Quantitative Analysis**: Sophia quantified the movement and localization in the compounds using image analysis software. This allowed her to generate quantitative data on the compounds' behavior, including their distribution, concentration, and movement dynamics within different cellular compartments.

Conclusion

Sophia's multidisciplinary approach not only solved the challenge but also paved the way for a deeper understanding of intracellular communication processes. Her innovative use of chromatography combined with microscopy provided a comprehensive view of how these labeled compounds contributed to cellular signaling, opening new avenues for therapeutic interventions. By integrating chromatography and microscopy, Sophia successfully tackled a complex scientific puzzle and made a significant contribution to the field of cellular biology.

Scenario-34

Emily, a dedicated Ph.D. student at a renowned research institute, has been working tirelessly on a project focused on understanding the role of a specific protein in a complex cellular process. As part of her research, she needs to quantify the levels of this protein across different experimental conditions. Emily has been using various techniques, but her current challenge involves **accurately quantifying the protein levels in a time-course experiment, where the protein's concentration changes dynamically**.

Scientific approach

Emily decided to utilize a combination of Spectroscopy and Blotting Techniques to address her protein quantification challenge. Here's how she devises her approach:

1. **Sample Preparation**: Emily carefully prepares samples from her experimental conditions, ensuring that the protein of interest is properly extracted and purified.

2. **Colorimetric Assay Selection**: She selects a colorimetric assay that is specific to the protein she's studying. This assay involves a reaction that generates a colored product in proportion to the protein concentration.

3. **Spectrophotometric Measurements**: Emily uses a UV-Vis spectrophotometer to measure the absorbance of the colored products generated by the colorimetric assay. By comparing the absorbance values of her samples against a standard curve created with known concentrations of the protein, she can determine the protein concentration in her experimental samples.

4. **Verification with Blotting Techniques**: To validate her results and ensure accuracy, Emily also performs Western blotting. She separates the proteins from her samples using gel electrophoresis and transfers them onto a membrane. She then uses antibodies that specifically recognize her protein of interest to visualize its presence on the blot.

5. **Densitometric Analysis**: Emily quantifies the protein bands on the Western blot using a densitometer or specialized software. By comparing intensity of the protein bands against known protein standards, she can confirm accuracy of her colorimetric assay-based measurements.

6. **Data Integration**: Emily integrates the data from both the colorimetric assay and Western blotting to refine her quantification. This integrated approach not

only provides precise protein concentration measurements but also validates the reliability of her results.

Conclusion

By combining Spectroscopy with Blotting Techniques, Emily successfully overcomes the challenge of quantifying dynamic changes in protein concentration. Her interdisciplinary approach showcases the power for integrating different methodologies to achieve robust and accurate scientific results.

Scenario-35

Meet Sarah, a dedicated PhD candidate at a prestigious research institute. Her passion lies in unraveling the intricate details of protein behavior within cells. She's been given a challenging project: **to comprehensively analyze a newly discovered protein, from its expression to its interactions with other molecules**.

Scientific challenge

The protein, named "CellularX," is suspected to play a crucial role in a specific signaling pathway. However, its expression level, cellular localization, and potential interactions with other proteins are largely unknown. Sarah's task is to devise a strategy that provides a deep understanding of CellularX's characteristics

Scientific approach

1. **Electrophoresis**: Sarah begins by employing electrophoresis to determine the molecular weight of CellularX. She uses SDS-PAGE to separate proteins based on size and runs a gel containing samples from various cell lines. By comparing the migration of CellularX with known protein markers, she estimates its molecular weight. This initial step helps her verify if the protein's size matches predictions.

2. **Immunotechniques**: Next, Sarah employs immunotechniques. She generates antibodies specific to CellularX using recombinant protein fragments. Western blotting helps her confirm the protein's expression levels across different cell types and under various conditions. By comparing the intensity on bands with a loading control, she can infer if CellularX's expression changes in response to specific stimuli.

3. **Molecular Techniques**: To understand CellularX's interactions, Sarah turns to molecular techniques. She performs co-immunoprecipitation (Co-IP) assays,

using antibodies against CellularX to pull down interacting partners. Mass spectrometry analysis of the pulled-down proteins reveals potential binding partners, shedding light on CellularX's role in the pathway.

Conclusion

Sarah's triple approach—Electrophoresis, Immunotechniques, and Molecular Techniques—provides a comprehensive view of CellularX. By combining SDS-PAGE for size determination, Western blotting for expression analysis, and Co-IP for interaction studies, she uncovers essential information about CellularX's role, expression patterns, and potential partners within the cellular environment. Through her dedication and the power of these techniques, Sarah contributes to the broader understanding of CellularX's involvement in the signaling pathway, opening new avenues for therapeutic interventions and further research.

Scenario-36

Emma, a passionate Ph.D. researcher at a prestigious biotechnology institute, is captivated by the intricate web of protein-protein interactions that govern cellular processes. Her curiosity leads her to tackle a challenging research problem involving the **validation of crucial protein interactions within a signaling pathway implicated in cancer progression**.

Scientific challenge

Emma's challenge revolves around identifying and validating protein-protein interactions that play a pivotal role in the pathway. This is no small feat since these interactions are transient and elusive, often occurring in the dynamic microenvironment of the cell. To unravel this complex conversation, Emma must pinpoint the interactions that are biologically relevant.

Scientific approach

1. **Yeast Two-Hybrid Assay (Molecular Technique)**: Emma decides to employ the yeast two-hybrid (Y2H) assay, a powerful molecular technique. In this approach, she designs two different constructs: one containing a DNA-binding domain fused to Protein A (Prey), and the other containing an activation domain fused to Protein B (Bait). If Protein A and Protein B interact in the yeast cell, they bring together the DNA-binding and activation domains, activating the reporter gene and allowing the yeast to grow in selective media.

Virtual learning aid

Scan It!

2. **Immunoprecipitation (Immunotechnique)**: she adds a layer of validation by utilizing immunoprecipitation. She genetically engineers the cells to express a tagged version of Protein A and Protein B. After performing the Y2H assay, Emma harvests the yeast cells and employs antibodies specific to the tags to immunoprecipitate the putative interacting proteins. This step helps confirm the interactions in a more native context.

3. **Mass Spectrometry (Analytical Technique)**: Emma employs mass spectrometry to identify other potential interactors in the complex. By comparing the interactions detected through Y2H and immunoprecipitation with the proteins identified by mass spectrometry, Emma gains a comprehensive view of the protein-protein interaction network.

Conclusion

Emma successfully navigates the intricate landscape of protein-protein interactions. Through her meticulously designed experiments, she confirms previously known interactions and uncovers novel ones. Her work sheds light on how this signaling pathway orchestrates critical cellular processes, making significant contributions to our understanding of cancer biology. In this scenario, Emma's scientific approach highlights the synergy between molecular techniques (Y2H) and immunotechniques (immunoprecipitation) along with analytical tools (mass spectrometry) to address a complex biological question.

Scenario-37

Emily, a dedicated Ph.D. student at a prestigious research institution, is embarking on an ambitious project involving enzyme kinetics. Her focus is on **understanding the catalytic activity of an enzyme responsible for a crucial metabolic process**. The enzyme is known to play a significant role in disease progression, making it an exciting target for therapeutic interventions.

Scientific challenge

Emily needs to determine the enzyme's kinetic parameters, including its Michaelis-Menten constant (Km) and maximum reaction rate (Vmax), which are essential for understanding its behavior. However, the enzyme's intricate properties and the complexity of the reaction make this a challenging task.

Scientific approach

Emily decided to use a combination of spectroscopic techniques and molecular assays to overcome the challenge. Her approach involves the following steps:

1. **Selection of Substrate and Assay Design**: Emily carefully selects a substrate that mimics the enzyme's natural substrate. She designs an assay that allows her to monitor the reaction's progress using spectroscopy. The substrate is chosen to produce a measurable spectral change upon enzyme catalysis.

2. **Experimental Setup**: Emily sets up her experiments, ensuring consistent conditions such as temperature, pH, and buffer composition. She prepares solutions of varying substrate concentrations, creating a range that covers both low and high concentrations.

3. **Spectroscopic Measurements**: Emily uses spectroscopic techniques such as UV-Vis absorption spectroscopy or fluorescence spectroscopy, depending on the nature of the enzyme and substrate. These techniques allow her to monitor changes in absorbance or fluorescence intensity as the reaction progresses.

4. **Data Analysis**: Emily collects time-dependent spectral data for each substrate concentration. She plots the changes in absorbance or fluorescence over time for each concentration, creating kinetic profiles.

5. **Determining Kinetic Parameters**: Using the kinetic profiles, Emily can determine the initial reaction rates for different substrate concentrations. She employs the Michaelis-Menten equation and performs nonlinear regression analysis to extract the enzyme's Km and Vmax values.

6. **Validation and Interpretation**: To ensure the accuracy of her results, Emily conducts control experiments and repeats the assays multiple times. She also compares her findings to existing literature to validate her enzyme kinetics data. With the Km and Vmax values in hand, she can now elucidate the enzyme's catalytic efficiency and behavior.

Conclusion

By combining spectroscopic techniques with meticulous experimental design and data analysis, Emily successfully overcomes the challenge of studying enzyme kinetics for this complex enzyme. Her comprehensive scientific approach paves the way for deeper insights into the enzyme's role in disease processes and potential therapeutic interventions.

Scenario-38

Anna is a dedicated PhD student at a prestigious research institute, passionate about deciphering the intricate world of post-translational modifications (PTMs) in proteins. Her current project involves **investigating a specific protein's phosphorylation status, a PTM crucial for its function**.

Scientific challenge

The protein of interest is suspected to undergo phosphorylation at multiple sites in response to cellular stress. Anna's challenge is to precisely map these phosphorylation sites and understand their impact on the protein's activity.

Scientific approach

1. **Protein Extraction and Immunoprecipitation**: Anna starts by isolating the protein from cellular lysates. She carefully designs an immunoprecipitation (IP) experiment using an antibody that specifically recognizes the phosphorylated form of the protein. This allows her to selectively pull down the phosphorylated protein from the mixture.

2. **Gel Electrophoresis and Blotting**: Next, Anna separates the immunoprecipitated proteins based on their size using SDS-PAGE. She then transfers the proteins onto a nitrocellulose membrane through Western blotting. This immobilizes the proteins in a spatial arrangement that maintains their separation according to molecular weight.

3. **Primary and Secondary Antibodies**: Anna employs a primary antibody that binds specifically to the phosphorylated sites on the protein. This primary antibody is tagged with a fluorescent marker, enabling detection. After washing away excess primary antibody, she applies a secondary antibody conjugated with an enzyme that catalyzes a color-producing reaction upon exposure to a substrate.

4. **Visualization and Analysis**: Anna visualizes the results using a gel documentation system. Bands corresponding to the phosphorylated protein at different sizes indicate multiple phosphorylation sites. The enzyme-substrate reaction produces color bands, the intensity of which correlates with the degree of phosphorylation at each site.

5. **Mass Spectrometry Validation**: To validate her findings, Anna plans to perform mass spectrometry on the phosphorylated protein bands. This technique will accurately identify the amino acid residues that undergo phosphorylation, confirming the PTM sites she detected.

6. **Bioinformatics and Functional Analysis**: With the confirmed PTM sites, Anna utilizes bioinformatics tools to predict the impact of phosphorylation on the protein's structure and function. This information guides her subsequent experiments to understand how these modifications influence cellular processes.

Conclusion

Anna's rigorous scientific approach involving immune techniques and blotting techniques not only identifies phosphorylation sites but also opens avenues for in-depth functional exploration. Through her meticulous research, she advances the understanding of PTMs and their role in cellular signaling pathways.

Scenario-39

Meet Sarah, a passionate graduate student at a prestigious botanical research institute. Her fascination with plants and their hidden chemical wonders led her to embark on a challenging research project. Sarah's task is **to purify and characterize a rare natural compound from a plant extract, which holds potential medicinal properties**. The compound is believed to have antioxidant properties, making it a possible candidate for future pharmaceuticals.

Scientific challenge

Sarah faces the complex task of isolating the specific natural compound from a mixture of various plant constituents. The plant extract contains a diverse range of compounds, each with different chemical properties and polarities. Traditional separation techniques are not sufficient to isolate the compound to a pure state.

Scientific approach

Sarah devises a comprehensive approach that involves both chromatography and spectroscopy to conquer this challenge.

1. **Chromatography Phase**: Sarah starts with column chromatography, utilizing a column packed with an adsorbent material that will interact differently with various compounds in the extract. She employs a solvent system with varying polarities to separate the mixture into distinct fractions.

2. **Fraction Analysis**: As fractions are collected from the column, Sarah uses thin-layer chromatography (TLC) to analyze them. TLC provides a quick visual assessment of each fraction's composition and purity.

3. **Spectroscopy Integration**: Sarah's key breakthrough comes when she identifies a promising fraction through TLC. She then employs mass

spectrometry (MS) to determine the molecular mass of the compound. This helps her narrow down potential structures.

4. **High-Performance Liquid Chromatography (HPLC)**: To further enhance the purity of the compound, Sarah employs preparative HPLC. This advanced technique allows her to separate compounds based on their interactions with a stationary phase and elution through a mobile phase. The pure compound is collected at a specific retention time.

5. **Spectroscopic Characterization**: With a purified compound in hand, Sarah performs nuclear magnetic resonance (NMR) spectroscopy to elucidate the compound's structure. NMR provides information about the connectivity on atoms within the molecule.

Conclusion

Through the integration of chromatography and spectroscopy, Sarah successfully isolates and characterizes the elusive natural compound from the plant extract. The compound's structure is confirmed using NMR, validating its potential as an antioxidant. Sarah's research contributes valuable insights to the field of natural product chemistry, and her findings open doors to further studies on its potential medicinal applications.

Scenario-40

Jane, a dedicated Ph.D. candidate at a prestigious research institute, is engrossed in her quest to decode the intricate world of epigenetics. Her focus is on understanding DNA methylation patterns and their impact on gene expression. As she delves deeper into her studies, she stumbles upon a unique challenge: **identifying and characterizing DNA methylation patterns in a specific gene associated with a rare hereditary disorder**.

Scientific challenge

The gene of interest, named "EpiGeneX," has been linked to the disorder, but its methylation status remains elusive. Jane's supervisor, impressed by her dedication, tasked her with designing an experimental approach to uncover the DNA methylation patterns within the "EpiGeneX" gene. This task is not only critical for advancing the field of epigenetics but also holds potential implications for understanding and potentially treating the disorder.

Scientific approach

Jane begins her scientific journey by outlining a comprehensive plan that integrates electrophoresis and molecular techniques to unravel the DNA methylation mystery.

1. **DNA Extraction and Bisulfite Conversion**: Jane starts by isolating DNA from both affected and healthy cells. She then employs bisulfite conversion, a technique that converts unmethylated cytosines to uracils while leaving methylated cytosines intact. This conversion is crucial for distinguishing methylated from unmethylated cytosines during subsequent analysis.

2. **PCR Amplification**: Utilizing bisulfite-converted DNA as a template, Jane performs PCR to selectively amplify the "EpiGeneX" region. She designs primers that target both methylated and unmethylated sequences.

3. **Agarose Gel Electrophoresis**: Jane employs agarose gel electrophoresis to separate the PCR products based on their size. The distinct bands obtained represent the amplified DNA fragments. By comparing the band patterns between affected and healthy samples, Jane gains insights into potential methylation differences.

4. **Bisulfite Sequencing**: To precisely determine the methylation status on individual cytosines within the "EpiGeneX" region, Jane turns to bisulfite sequencing. She clones the PCR products into vectors and sequences multiple clones. By analyzing the sequencing results, she pinpoints the exact cytosines that are methylated or unmethylated.

5. **Data Integration**: Jane integrates the results from agarose gel electrophoresis and bisulfite sequencing. This integration enables her to correlate the observed band patterns with specific methylation patterns, providing a comprehensive view of the "EpiGeneX" gene's methylation landscape.

Conclusion

Jane's meticulous approach not only uncovers the DNA methylation patterns within the "EpiGeneX" gene but also establishes a robust methodology that can be applied to other genes and disorders in the realm of epigenetics. Her dedication and innovative thinking propel the field forward, bringing us one step closer to understanding the delicate interplay between DNA methylation and human health.

Scenario-41

Meet Sarah, a dedicated Ph.D. student at a renowned research institute. Her passion lies in understanding how cells behave within three-dimensional (3D) environments, which more accurately mimic the conditions found in living organisms. Her challenge is **to study the growth patterns and interactions in cancer cells within a 3D matrix, as opposed to the traditional two-dimensional (2D) cell culture**. This is crucial because cellular responses in 3D environments can differ significantly from those in flat, 2D cultures.

Scientific approach

1. **3D Cell Culture Preparation**: Sarah prepares a 3D culture system that closely mimics the extracellular matrix of a living organism. She uses a hydrogel matrix enriched with nutrients to support cell growth.

2. **Cell Seeding**: Sarah carefully seeds the cancer cells into the 3D matrix. She ensures uniform distribution and allows the cells to settle within the matrix, imitating the natural conditions.

3. **Live Cell Imaging**: Using confocal microscopy, Sarah captures live images of the 3D cell culture over time. This technique provides high-resolution images of cells within the matrix, allowing her to track their movements, interactions, and growth dynamics in real time.

4. **Staining Techniques**: Sarah employs fluorescent staining to differentiate between different cell types and structures within the 3D culture. This helps her study cell behavior and identify changes, such as proliferation, migration, and apoptosis.

5. **Image Analysis**: She utilizes specialized software to analyze the acquired images. This software allows her to reconstruct 3D images from the collected data, providing insights into cell distribution and spatial relationships.

6. **Quantitative Analysis**: Sarah quantifies parameters like cell density, cell-cell interactions, and the presence of cellular structures. This data helps her draw meaningful conclusions about how the 3D environment influences cancer cell behavior.

7. **Comparison with 2D Cultures**: To highlight the significance of her findings, Sarah performs parallel experiments using traditional 2D cell cultures. This comparative analysis helps her pinpoint differences in cell behavior between the two conditions.

8. **Result Interpretation**: Sarah interprets her data to understand how cancer cells respond to 3D environments, shedding light on potential differences in drug resistance, metastatic potential, and other critical factors.

Conclusion

By addressing the challenges of studying cells in 3D cultures, Sarah's research contributes to our understanding of cancer biology and could pave the way for more effective drug testing and treatment strategies. Sarah's journey showcases the vital role that combining microscopy with 3D cell culture plays in advancing our understanding of cellular behavior within complex environments.

Bibliography

Abbas AK, Lichtman AH and Pillai S (2010), Cellular and Molecular Immunology, 6th ed. Saunders

Alberts B et al (2008), Molecular Biology of the Cell, 5th ed. Garland Science Publishing.

Bai XC, Yan C, Yang G et al (2015), An atomic structure of human y-secretase, Nature, 525, 212.

Bakker E (2004), Electrochemical sensors, Analytical Chemistry 2004, 76 (12), 3285-3298.

Banwell CN, McCash EM (2013), Fundamentals of Molecular Spectroscopy, 5th ed. McGraw Hill Education.

Becker WM, Kleinsmith LJ and Hardin J (2006), The world of the cell, 6th ed. Pearson Education Inc.

Behrmann E, Loerke J, Budkevich TV et al (2015), Structural snapshots of actively translating human ribosomes, Cell, 161, 845.

Berg JM, Tymoczko JL and Stryer L (2006), Biochemistry, 6th ed. W.H. Freeman and Company.

Berg JM, Tymoczko JL and Stryer L (2007), Biochemistry, 6th ed. W.H. Freeman and Company.

Berk AJ (1989), Characterization of RNA molecules by S1 nuclease analysis. Methods in Enzymology, 180, 334-347

Bozzola JJ and Russell LD (1998), Electron Microscopy: Principles and Techniques for Biologists, 2nd ed. Jones and Bartlett Publishers, Inc.

Brown TA (2007), Genomes, 3rd ed. Garland Science Publishing.

Clark BJ (1993), UV Spectroscopy: Techniques, Instrumentation and Data Handling, Chapman and Hall Inc.

Coico R, Sunshine G and Benjamini E (2003), Immunology, A short course, 5th ed. John Wiley and Sons, Inc.

Decker J and Reischl U (2004), Molecular diagnosis of infectious diseases, 2nd ed. Humana Press.

Divan A and Royds J (2013), Tools and techniques in biomolecular science, Oxford University Press.

Dorsey JG, Foley JP, Cooper WT, Barford RA and Barth HG (1990), Liquid chromatography: theory and methodology. Analytical Chemistry 62:324R-356R.

Eggins B (2002) Chemical sensors and biosensors, Analytical Techniques in the Sciences. John Wiley & Sons, West Sussex.

Erlich HA. Gelfand D and Sninsky JJ (1991), Recent advances in the polymerase chain reaction. Sciences, 252: 1643-1651.

Fields S and Sternglanz R (1994), The two hybrid system: an assay for protein-protein interactions, Trends in Genetics, 10, 286-292.

Fields, S and Song OK (1989), A novel genetic system to detect protein-protein interactions. Nature, 340: 245-246.

Freeman WM, Walker SJ and Vrana KE (1999), Quantitative RT-PCR: Pitfalls and potential. BioTechniques, 26: 112-125.

Freifelder D (1982), Physical Biochemistry, 2nd ed. W.H. Freeman and Company.

Garfin DE (2003), Two-Dimensional Gel Electrophoresis: An Overview. Trends in Analytical Chemistry 22 263-272.

Gilmartin PM and Bowler DChris (2002), Molecular Plant Biology: A Practical Approach, Volume 2, Oxford university press.

Hames BD (1998), Gel Electrophoresis of Proteins: A Practical Approach, 3rd ed. Oxford University Press, New York.

Harrison RG, Todd, Paul, Rudge, Scott R and Petrides DP (2003), Bioseparations - Science and Engineering. Oxford University Press, 2003.

Hollas JM (2002), Basic Atomic and Molecular Spectroscopy, RSC Publishing.

Holler FJ, Crouch SR (2014), Fundamentals of analytical chemistry, 9th ed. Cengage Learning.

Hoppert M (2003), Microscopic Techniques in Biotechnology, Wiley-VCH Verlag.

Jorgenson JW (1986), Electrophoresis. Analytical Chemistry 58:743A-760A.

Keeler J (2011), Understanding NMR Spectroscopy, 2nd ed. John Wiley and Sons.

Khudyakov YE and Fields HA (2010), Artificial DNA: Methods and Applications, CRC Press.

Lakowicz JR (1999), Principles of Fluorescence Spectroscopy, 2nd ed. Kluwer, New York.

Law B (2005), Immunoassay: A practical guide, Taylor and Francis.

Lodish H, Berk A, Kaiser C et al (2007), Molecular Biology of the Cell, 4th ed. W.H. Freeman and Company.

Lodish H, Berk A, Kaiser CA, et al (2008), Molecular Cell Biology, 6th ed. New York: WH Freeman.

Manz A, Pamme N and Dimitri L (2004), Bioanalytical Chemistry, 3rd ed. Imperial College Press.

McGraw Hill. Westermeier R (2005), Electrophoresis in Practice, 4th ed. Wiley VCH.

McHale JL (2017), Molecular spectroscopy, 2nd edition. CRC press.

McPherson M and Moller S (2006), PCR - The basics, 2nd ed. Taylor and Francis group.

Merk A, Bartesaghi A, Banerjee S et al, (2016), Breaking Cryo-EM Resolution Barriers to Facilitate Drug Discovery, Cell 165, 1698.

Miesfeld RL (1999), Applied Molecular Genetics, Wiley-Liss, Inc.

Miller JM (2005), Chromatography: Concepts and Contrasts, 2nd ed. John Wiley and Sons Inc.

Nicholl ST (2002), An Introduction to Genetic Engineering, 3rd ed. Cambridge University Press.

Owen J, Punt J and Stranford S (2013), Kuby Immunology, 7th ed. Macmillan Publishers.

Parra I and Windle B (1993), High resolution visual mapping of stretched DNA by fluorescent hybridization. Nature Genetics 5, 17-21.

Paul WE (2008), Fundamental immunology, 6th ed. Philadelphia, Wolters Kluwer/Lippincott Williams & Wilkins.

Pavia DL, Lampman GM, Kriz GS and Vyvyan JR (2009), Introduction to spectroscopy, 4th ed. Brooks/Cole Cengage Learning.

Pawley J (2006), Handbook of Biological Confocal Microscopy, 3rd ed. Springer.

Pergande MR and Cologna SM (2017), Isoelectric Point Separations of Peptides and Proteins, Proteomes.

Ried T, Schröck E, Ning Y, et al. (1998), Chromosome painting: a useful art. Hum Mol Genet 7:1619-1626.

Roitt I, Brostoff J and Male D (2012), Immunology, 8th ed. Saunders Elsevier.

Ronaghi M, Ehleen M and Nyrn P (1998), A sequencing method based on real-time pyrophosphate. Science, 281, 363-365.

Sirohi D, Chen Z, Sun L, et al (2016), The 3.8Å resolution cryo-EM structure of Zika Virus, Science, 352, 467.

Skoog DA, Holler FJ and Crouch SR (2007), Principles of instrumental analysis, 6th ed. Thomson Publishing, USA.

Speicher MR and Carter NP (2005), The new cytogenetics: Blurring the boundaries with molecular biology. Nature Reviews Genetics 6, 782-792.

Stanley J (2002), Essentials of immunology and serology, Cengage Learning.

Stoughton RB (2005), Applications of DNA microarrays in biology. Annu Rev Biochem, 74:53-82.

Strachan T and Read AP (2004), Human Molecular Genetics, 3rd ed. Garland Publishing.

Thiel T, Bissen S and Lyons E (2002), Biotechnology: DNA to Protein; A Laboratory Project.

VanGuilder HD, Vrana KE and Freeman WM (2008), Twenty-five years of quantitative PCR for gene expression analysis. Biotechniques, 44, 619-624.

Weissensteiner T, Nolan T, Bustin SA, Griffin HG and Griffin A (2004), PCR Technology: Current Innovations, 2nd ed. CRC Press.

Wilson K and Walker JM (2000), Principles and Techniques of Practical Biochemistry, 5th ed. Cambridge University Press.

Dear readers, I am Suryaprakash Tripathy, the author of "Demystifying Biotechnology". The inspiration behind this book springs from a personal journey. During my postgraduate studies, I encountered the intricate world of biotechnology and came to understand the pressing need for a tool that could simplify and demystify its complex techniques. Throughout my master's program, I grappled with the challenges of time and space constraints while trying to grasp these intricate processes. I yearned for a guiding presence, a mentor in the form of a comprehensive resource that would house all the protocols, concepts, and insights in one place. This resource would not only facilitate my learning but also empower me to design experiments and contribute meaningfully to the biotechnology community. I envisioned having a companion who could nurture my knowledge at my own pace, in my own space. With the aspirations of countless biotechnology enthusiasts in mind, I have crafted this book as a futuristic, personal tool for your success in the biotechnology realm. I invite you to stay connected with the author and share your valuable insights. Your feedback and suggestions are not only welcome but highly encouraged. Your input is invaluable, and I look forward to our journey of learning and growth together.

Email Id	
LinkedIn	

Dear readers, this book is not just a part of what I do; it is a part of who I am. I truly believe that I cannot become the best version of myself without your valuable input. I would be immensely grateful if you could take a moment to share a priceless review on the Amazon page from where you purchased this book. I extend my gratitude for joining me on this journey and for your invaluable support.

9 789360 136659